Supercapacitors

Concepts and advances

Online at: https://doi.org/10.1088/978-0-7503-5944-3

Supercapacitors

Concepts and advances

Abha Misra

Department of Instrumentation and Applied Physics, Indian Institute of Science, Bangalore 560012, India

IOP Publishing, Bristol, UK

ISBN 978-0-7503-5944-3 (ebook)
ISBN 978-0-7503-5942-9 (print)
ISBN 978-0-7503-5945-0 (myPrint)
ISBN 978-0-7503-5943-6 (mobi)

DOI 10.1088/978-0-7503-5944-3

Version: 20250901

IOP ebooks

British Library Cataloguing-in-Publication Data: A catalogue record for this book is available from the British Library.

Published by IOP Publishing, wholly owned by The Institute of Physics, London

IOP Publishing, No.2 The Distillery, Glassfields, Avon Street, Bristol, BS2 0GR, UK

US Office: IOP Publishing, Inc., 190 North Independence Mall West, Suite 601, Philadelphia, PA 19106, USA

Contents

Preface ix

Acknowledgments x

Author biography xi

1 Concept of the electrochemical double layer **1-1**

1.1 Thermodynamics of double layers 1-1

1.2 Excess charge and capacitance 1-4

1.3 Models for the formation of a double layer 1-5

1.3.1 The Helmholtz model 1-5

1.3.2 The Gouy–Chapman theory 1-6

1.3.3 Stern's modification 1-9

1.4 Double layers in solids 1-11

1.5 Single crystal electrode surfaces 1-12

1.6 The solid metal–solution interface 1-12

1.7 Rate of specific adsorption 1-12

1.8 Effect of the double layer on electrode reaction rates 1-13

1.9 Double layers in nonaqueous and aqueous electrolytes 1-14

1.9.1 Aqueous media 1-17

1.9.2 Nonaqueous media 1-18

References 1-18

2 Concepts of electric capacitors and electrolytic capacitors **2-1**

2.1 Capacitance, electric field, and potential 2-1

2.1.1 Energy storage efficiency 2-2

2.1.2 Dielectric breakdown field 2-3

2.1.3 Discharge time 2-3

2.2 Polarization mechanisms 2-3

2.3 Dielectric materials 2-4

2.3.1 Ceramic-based dielectrics 2-5

2.3.2 Polymer- and glass-based dielectrics 2-7

2.4 Types of capacitors 2-7

2.4.1 Linear and nonlinear capacitors 2-7

2.4.2 Ferroelectric capacitors 2-8

2.4.3 Polar polymer dielectric capacitors 2-8

2.4.4 Fixed and variable capacitors 2-8

2.4.5 Power capacitors 2-8
2.4.6 High-voltage capacitors 2-8
2.4.7 Interference-suppression capacitors 2-8
2.5 Charging and discharging of capacitors 2-9
2.6 Energy storage mechanism and electrical circuits 2-9
2.7 Types of electrolytic capacitors 2-11
References 2-12

3 Concepts of electrochemical processes 3-1

3.1 Introduction and overview of electrode processes 3-1
3.2 Kinetics of electrode processes 3-2
3.3 Basic potential methods 3-4
3.4 Electrochemical instrumentation 3-5
3.5 Cyclic voltammetry 3-6
3.6 Galvanostatic charge–discharge analysis 3-12
3.7 AC impedance analysis 3-16
References 3-21

4 Supercapacitor concepts 4-1

4.1 Overview 4-1
4.2 Electric double-layer capacitance and pseudocapacitance 4-4
4.2.1 Charge storage mechanism in electric double-layer capacitors 4-4
4.2.2 Pesudocapacitance 4-5
4.3 Supercapacitor design, fabrication, operation, and experimental parameters 4-13
4.4 Equivalent series resistance 4-18
4.5 Leakage resistance 4-18
4.6 Supercapacitor charging and discharging 4-19
4.7 Energy and power densities 4-21
4.8 Concepts of batteries and supercapacitor energy storage 4-22
4.9 Comparison of batteries and supercapacitors 4-23
4.10 Density of charge storage 4-23
4.11 Comparison of charge and discharge behaviors in supercapacitors and batteries 4-23
4.12 Supercapacitor cell balancing 4-26
4.13 Supercapacitor manufacture and construction 4-26
4.14 Series configuration of supercapacitors 4-26

4.15 Parallel configuration of supercapacitors 4-27
4.16 Methods of balancing 4-27
References and further reading 4-27

5 Materials for supercapacitors 5-1
5.1 Metal oxides 5-1
5.1.1 Manganese oxide (MnO_2) 5-1
5.1.2 Ruthenium oxide (RuO_2) 5-2
5.2 Metal hydroxides 5-3
5.3 Carbon 5-3
5.4 Polymers 5-6
5.5 Electrolytes 5-7
5.6 Separators 5-9
5.7 Current collectors 5-9
References and further reading 5-10

6 Hybrid supercapacitors 6-1
References and further reading 6-14

7 Microsupercapacitors 7-1
7.1 Device architecture design 7-2
7.2 Device charge storage mechanism 7-3
7.3 Microsupercapacitors compared to microbatteries 7-6
7.4 Microsupercapacitor fabrication methods 7-7
7.5 Electrochemical measurements 7-7
7.6 Devices in parallel and series configurations 7-8
7.7 Advantages of 2D and 3D electrodes 7-8
7.8 Electrolytes for microsupercapacitors 7-9
References and further reading 7-11

8 Self-powered supercapacitors and batteries 8-1
8.1 Self-powered supercapacitors 8-1
8.2 Triboelectric nanogenerator supercapacitors 8-1
8.3 Piezoelectric supercapacitors 8-2
8.4 Thermoelectric supercapacitors 8-5
8.5 Solar energy or light-based supercapacitors 8-7

8.6 Moisture-based self-charging in supercapacitors 8-10
8.7 Optically sensitive electrochemical reactions 8-11
References 8-14

9 Applications of supercapacitor 9-1

9.1 Electrical vehicles 9-1
9.2 Integrated EV systems with supercapacitors and a battery pack 9-1
9.3 Regenerative braking 9-2
9.4 Start-stop systems 9-3
9.5 Uninterruptible power supplies 9-4
9.6 Microgrid applications 9-4
9.7 Portable applications 9-5
9.8 Supercapacitors in telecommunications 9-6
9.9 Cold-start behavior of diesel-fueled engines 9-6
9.10 Applications of microsupercapacitors 9-6
9.10.1 Energy storage devices 9-6
9.10.2 Self-powered wireless sensors 9-7
9.10.3 Stimuli-responsive devices 9-7
9.10.4 Medical diagnostics 9-8
9.10.5 AC line filtering 9-9
References 9-9

Preface

The idea of this book is to provide readers who are researchers, undergraduates, and postgraduate students with a comprehensive understanding of supercapacitors. The fundamentals of electrochemistry will help the reader understand the importance of various components of supercapacitors and their selection in attaining optimal performance. Capacitors in other forms are also described for contrast and to relate the differences in the materials that are used to make them. This book helps the reader distinguish between electric double-layer capacitance and pseudocapacitance using their electrochemical responses. Advances in supercapacitors to hybrid supercapacitors, microsupercapacitors, and self-powered supercapacitors are discussed, with examples that will motivate researchers in the field to take advantage of a deep understanding of supercapacitor devices. This book promotes innovation in developing novel supercapacitor devices that achieve impressive energy density. The development of ultrahigh energy density in on-chip microsupercapacitors is expected to occur in the near future and requires the incorporation of novel ideas gained from fundamental learning.

All the chapters provide unique, comprehensive details of electrochemical energy storage, starting from the electrochemical interfaces between the electrode and electrolyte to advanced applications. The fundamental formulation of electrochemistry in the context of the formation of the electric double layer is adopted from the book by Allen J Bard *et al*, and also Conway's basics are used to connect it with different systems. Self-powered and micro-supercapacitors have gained much attention due to the idea of reduced power consumption and miniaturized energy storage systems. The chapters provide a collection of knowledge from various articles and book chapters to provide numerous examples of each case study. The concept of electric double layer is described in detail to set a context by differentiating from the electrostatic and electrolytic capacitors. The electrochemical process is also detailed by introducing kinetics and instrumentation. Supercapacitors consider both the pseudocapacitance and electric double layer capacitance. It is important for students and new researchers in the field to understand the differences in the analysis. Thus, it explains the choice of materials with many example responses and properties in a tabulation form.

The author has worked extensively in developing microsupercapacitors and has developed self-powering capability using optical illumination on electrodes made of heterointerfaces of various bandgap materials. The future research requires advanced interfaces and the design of planar electrodes to maximize the energy density. Transistor-based electrodes provide a novel supercapacitor to develop an on-chip energy storage system, which is not possible with a battery system. The book motivates researchers to bring innovation to the field. For easy reading, acronyms are defined repeatedly to maintain the connectivity with the different contexts.

Readers will find an easy read of the book to connect their results, analysis and conclusions. The method of analysis of cyclic voltammetry (CD), galvanostatic charge/discharge (GCD), and electrochemical impedance spectroscopy (EIS) results is detailed with various published examples.

Acknowledgments

I would like to acknowledge my students and postdoctoral researchers who worked with me to develop a deep understanding of on-chip microsupercapacitors and self-powered supercapacitors, which enabled me to develop the concept of this book.

Author biography

Abha Misra

Dr Abha Misra is a Professor in the Department of Instrumentation and Applied Physics, at Indian Institute of Science (IISc). She obtained her PhD from Indian Institute of Technology Bombay (IIT Bombay). Thereafter, she received Gordon and Betty Moore Foundation Postdoctoral Fellowship in California Institute of Technology (Caltech), USA. Dr Misra is an Associate of Indian Academy of Sciences, member of National Science Academy and TWAS Young Affiliate. She is also a recipient of INSA Medal for Young Scientists, SERB Women Excellence Award and SERB-POWER Fellowship. She also won BPCL Innovation Award under net zero emission category. Dr Misra's research at IISc is mainly focusing on developing devices for the self-powered energy storage and sensing applications. She serves as an editorial board member of *ACS Applied Materials and Interfaces* and an associate editor of *RSC Advances*. She has published more than 100 international research papers and authored two books.

IOP Publishing

Supercapacitors
Concepts and advances
Abha Misra

Chapter 1

Concept of the electrochemical double layer

The concept of electrical double layer is described theoretically in different models as two layers of opposing charges separated by an atomic-scale distance. Von Helmholtz proposed the distribution of opposite charges in quasi two-dimensional planes, which was first used to describe the interfaces of colloidal particles. In colloidal interfaces, the charges on the electrode surface of the double layer can originate from either the ionization of an acid–base pair or the adsorption of ions. To balance the charge on the colloid, counterions of opposite charge accumulate on the solution side to form a double layer of negative and positive charges in an array [1].

The electrical double layer is formed from layers of positive and negative charges as well as a region of high electric field between the layers. It thus dominates in controlling the electrical and chemical properties of the surface. Three distinct double layers, namely, the inner Helmholtz plane (IHP), the outer Helmholtz plane (OHP) (as shown in figure 1.1), and the diffuse layer form at the solid/liquid interface. The thermodynamics of double layer provides information about the way in which its properties change with potential and the modifications of the electrolyte due to the presence of different species [2, 3].

1.1 Thermodynamics of double layers

To understand the thermodynamics of double layers, we assume an interface of area A (an interfacial zone a few angstroms thick), which separates two phases, α and β, as shown in figure 1.2. In general, the Gibbs adsorption isotherm describes the properties of electrochemical interfaces. The electrochemical free energy of a reference system depends on the pressure, temperature, and molar quantities of all the components as follows: $G^R = G^R(P, T, n_i^R)$, where n_i^R is the number of moles of species i in the reference system. Here, the surface area has no role to play because the two phases have no interaction with the interface. It is important to describe the

doi:10.1088/978-0-7503-5944-3ch1

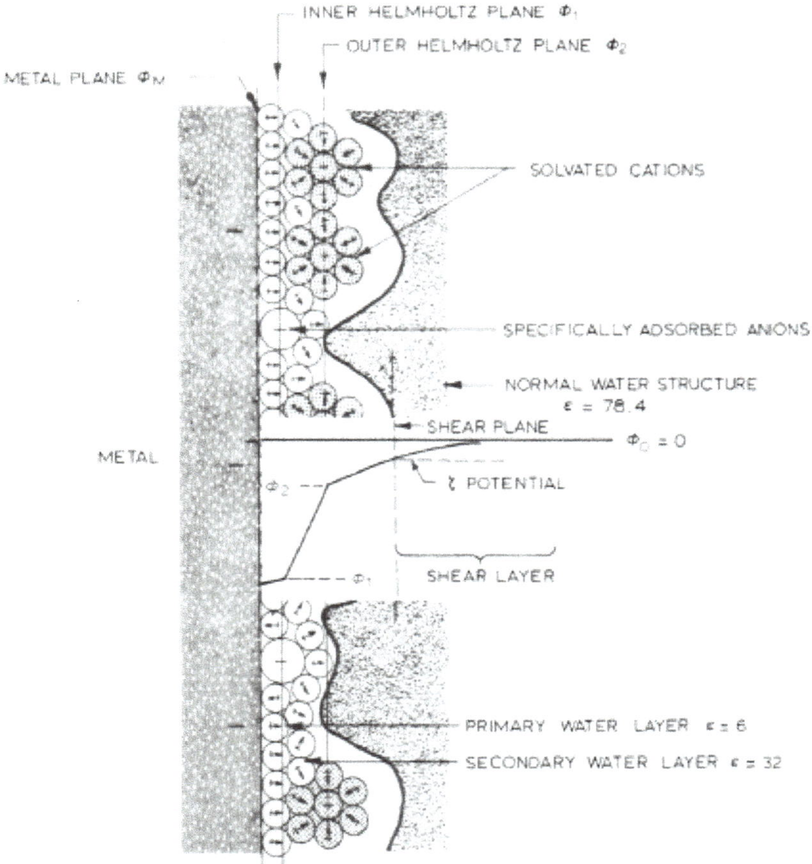

Figure 1.1. Double layer model at a metal–electrode interface. Reprinted from [4], Copyright (1983), with permission from Elsevier.

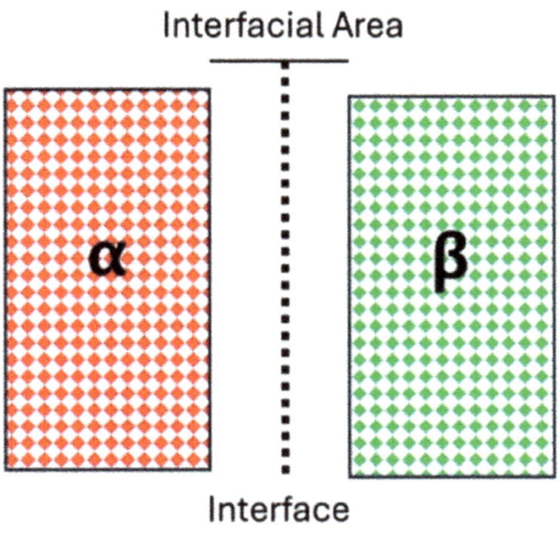

Figure 1.2. A schematic diagram showing the interfacial zone dividing two phases α and β.

surface excess quantity, which is relevant to the properties of the interface. It is mainly controlled by excesses and deficiencies in the concentrations of components [5, 6].

Thus, in any species, the surface excess in the number of moles is $n_i^{\sigma} = n_i^{S} - n_i^{R}$, where n_i^{σ} is the excess quantity and n_i^{S}, n_i^{R} are the number of moles of species i for the actual system and the reference system, respectively. Moreover, the interfacial area in real systems minimizes or maximizes; hence, the free energy, G^S, of the actual system depends on the area A as follows: $G^S = G^S(T, P, A, n_i^S)$. Thus, the differentials for the Gibbs energies of real and actual systems can be written as

$$dG^R = \left(\frac{\partial G^R}{\partial T}\right)dT + \left(\frac{\partial G^R}{\partial P}\right)dP + \sum_i \left(\frac{\partial G^R}{\partial n_i^R}\right)dn_i^R$$

$$dG^S = \left(\frac{\partial G^S}{\partial T}\right)dT + \left(\frac{\partial G^S}{\partial P}\right)dP + \left(\frac{\partial G^S}{\partial A}\right)dA + \sum_i \left(\frac{\partial G^S}{\partial n_i^S}\right)dn_i^S.$$

The electrochemical potential μ_i at constant pressure and temperature (given by the partial derivative $\partial G^R/\partial n_i^R$) remains constant at equilibrium. Thus,

$$\mu_i = \left(\frac{\partial G^R}{\partial n_i^R}\right) = \left(\frac{\partial G^s}{\partial n_i^s}\right).$$

Moreover, the partial derivative of the Gibbs energy with respect to area ($\frac{\partial G^R}{\partial A}$) is known as surface tension, γ, which is the energy required to form a unit area of new surface. The differential excess free energy can be written as

$$dG^{\sigma} = dG^S - dG^R = \gamma dA + \sum_i \mu_i d\left(n_i^s - n_i^R\right)$$

or

$$dG^{\sigma} = \gamma dA + \sum_i \mu_i dn_i^{\sigma}.$$

Further, using Euler's theorem, the interfacial free energy depends on two variables, A and n_i, as follows:

$$G^{\sigma} = \left(\frac{\partial G^{\sigma}}{\partial T}\right)A + \sum_i \left(\frac{\partial G^{\sigma}}{\partial n_i^{\sigma}}\right)n_i^{\sigma}$$

$$G^{\sigma} = \gamma A + \sum_i \mu_i n_i^{\sigma},$$

and the differentiation of the above equation is given by

$$dG^{\sigma} = \gamma A + \sum_i \mu_i n_i^{\sigma} + A d\gamma + \sum_i n_i^{\sigma} d\mu_i.$$

Equating the two differential terms of excess free energy, we obtain

$$Ad\gamma + \sum_i n_i^\sigma d\mu_i = 0$$

$$-d\gamma = \sum_i \Gamma_i d\mu_i.$$

This expression is the Gibbs adsorption isotherm, where Γ_i is the surface excess concentration, given by $\Gamma_i = n_i^\sigma / A$. This expression relates experimentally significant quantities, such as interfacial structure, with the surface tension in the double-layer structure. It also provides a basis of the electrocapillary equation; for example, for a cell containing Cu′Ag/AgCl/K^+, Cl^-, M/Hg/Ni/Cu, the equation is as follows,

$$-d\gamma = \sigma^M dE_- + \Gamma_{K^+(H_2O)} d\mu_{KCl} + \Gamma_{M(H_2O)} d\mu_M,$$

where E_- is the potential of the mercury electrode with respect to the reference, and σ^M is the excess charge density on the metallic side of the interface. This equation is known as the electrocapillary equation [7].

1.2 Excess charge and capacitance

Again, for a specific system as discussed earlier, a lifetime plot of a drop vs. potential (similar to the electrocapillary curve) exhibits a feature that reveals the presence of a maximum in surface tension at the potential of zero charge (PZC), where the excess charge on the electrode is given by

$$\sigma^M = -\left(\frac{\partial \gamma}{\partial E_-}\right), \text{ at pzc}, \quad \sigma^M = \sigma^S = 0.$$

The capacitance at the interface has the ability to store charge due to a perturbation in potential; thus, a small change in charge density due to a variation in the potential is defined as differential capacitance:

$$C_d = \left(\frac{\partial \sigma^M}{\partial E}\right).$$

Thus, the differential capacitance is the slope of the plot of σ^M vs. E, an example of which is shown in figure 1.3.

For any ideal capacitor, C_d does not remain constant as the potential changes. C_i is an integral capacitance, which is the ratio of σ^M (total charge density) at a potential E to the total potential difference, as given below [9]:

$$C_i = \frac{\sigma^M}{(E - E_z)},$$

where E_z is the PZC. Thus, C_i can be written as an average of C_d over the potential range from E_z to E as follows:

$$C_i = \frac{\int_{E_z}^{E} C_d dE}{\int_{E_z}^{E} dE}.$$

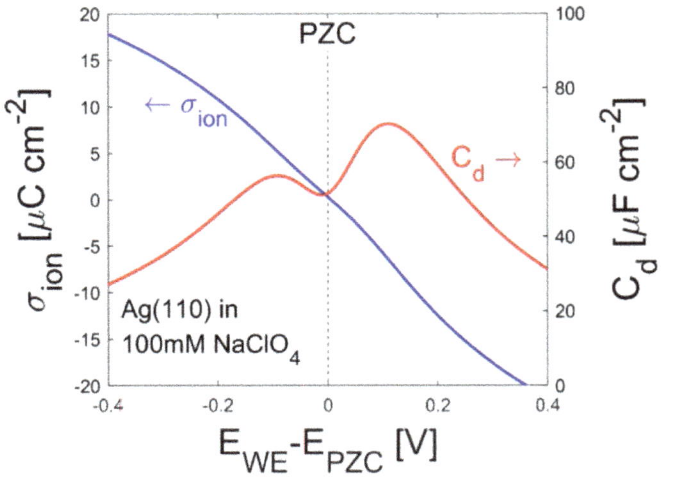

Figure 1.3. Schematic illustrating charge density vs. potential for integral and differential capacitances. Reproduced from [8]. CC BY 4.0.

1.3 Models for the formation of a double layer

In metals, the high free electron density at the surface is highly localized over a distance of 0.05–0.2 nm due to strong charge screening. This distance is called the Thomas–Fermi screening distance. However, in p- and n-type semiconductors, the charge carrier densities are smaller than those in metals by a factor varying from 10^{-4} to 10^{-15}. Thus, the charge distribution spreads away from the interface. The distribution of charge carriers extends into the bulk of the semiconductor, which varies inversely with the charge carrier density. Within semiconductors, the distribution of charge carriers near the interface has the same mathematical form as that of the ion distribution in the diffuse layer (on the solution side of the interface) [5, 10].

1.3.1 The Helmholtz model

Helmholtz proposed that the counter-charge in solution also stays at the electrode surface, forming two layers of oppositely polarized charges. These charges are separated by a small distance on the molecular scale. Thus, Helmholtz called it an electric double layer, equivalent to a parallel plate capacitor. If the voltage drop between the plates is V, then the stored charge density, σ, is given by $\sigma = \varepsilon\varepsilon_0 V/d$, where ε is the dielectric constant of the medium, ε_0 is the permittivity of free space, and d is the spacing between the plates. The differential capacitance is $d\sigma/dV = C_d = \varepsilon\varepsilon_0 V/d$, where C_d is a constant quantity. However, a dramatic capacitance response for different electrolyte concentrations is shown for an interface between mercury and sodium fluoride (NaF) in figure 1.4. This suggests that

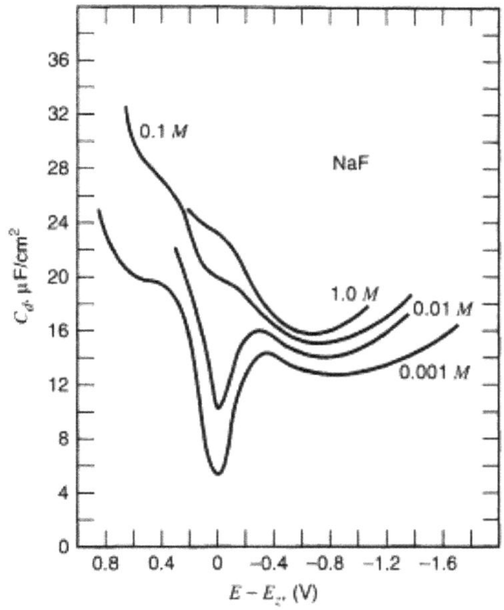

Figure 1.4. Variation of differential capacitance vs. potential for a NaF solution in contact with mercury. Reprinted with permission from [11]. Copyright (1947) American Chemical Society.

variations in C_d with potential and concentration could be due to ε or d, which also depend on variables such as concentration [2].

1.3.2 The Gouy–Chapman theory

Gouy and Chapman proposed the concept of a diffuse layer of charge in the electrolyte. The highly charged electrode has a compact diffuse layer; thus, C_d should be larger. The role of higher and lower concentrations of electrolyte is to cause stronger and weaker interactions due to the electrostatic forces, and eventually, the average distance of charge separation shows a dependence on potential and concentration. If the bulk concentration considered far from the electrode is n_i^0, then the population near the electrode is given by

$$n_i = n_i^0 \exp\left(\frac{-z_i e \varnothing}{kT}\right),$$

where $\varnothing$ is the potential with respect to bulk solution, e is the charge on the electron, k is the Boltzmann constant, T is the absolute temperature, and Z_i is the charge on ion i. The total charge per unit volume is

$$\begin{aligned}\rho(x) &= \sum_i n_i z_i e \\ &= \sum_i n_i^0 z_i e \exp\left(\frac{-z_i e \varnothing}{kT}\right),\end{aligned}$$

where i includes all the ion species present in the solution. Further, the Poisson equation also gives the charge density as a function of potential and distance, x, from the electrode:

$$\rho(x) = -\varepsilon\varepsilon_0 \frac{d^2\varnothing}{dx^2}.$$

Thus, the two equations yield

$$\left(\frac{d\varnothing}{dx}\right)^2 = \frac{2kT}{\varepsilon\varepsilon_0}\sum_i n_i^0\left[\exp\left(\frac{-z_i e\varnothing}{kT}\right) - 1\right].$$

The constants are evaluated by assuming that far from the electrode, $\varnothing = 0$ and $d\varnothing/dx = 0$.

Rearranging the equation gives the potential profile in the diffuse layer,

$$\int_{\varnothing_0}^{\varnothing} \frac{d\varnothing}{\sinh\,(ze\varnothing/2kT)} = -\left(\frac{8kTn^0}{\varepsilon\varepsilon_0}\right)^{1/2}\int_0^x dx,$$

where $\varnothing_0$ is the potential at $x = 0$, or the potential drop across the diffuse layer. The result is

$$\frac{\tanh ze\varnothing/4kT}{\tanh ze\varnothing_0/4kT} = e^{-\Re x},$$

where

$$\Re = \left(\frac{2n^0z^2e^2}{\varepsilon\varepsilon_0 kT}\right)^{1/2}.$$

Figure 1.5 shows a representative potential profile in a general way. The potential decays away from the surface. It is evident that at large potential, the diffuse layer is compact and gradually declines for smaller potential.

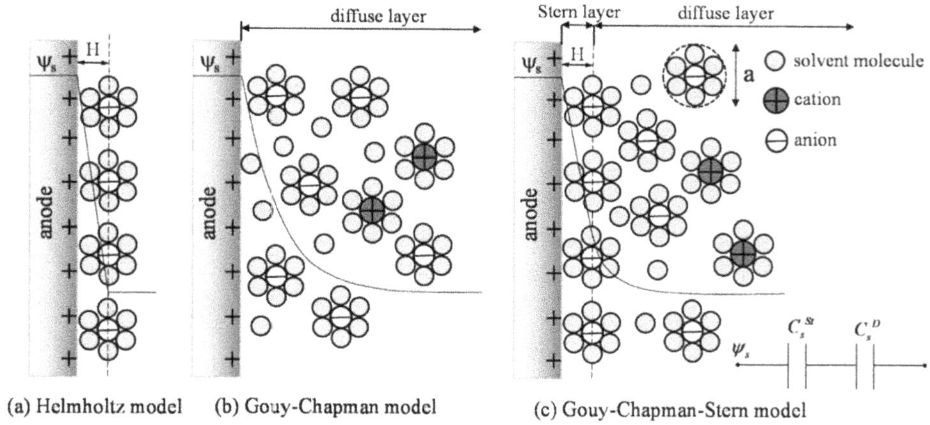

Figure 1.5. Schematic representation of the Gouy–Chapman model, depicting the potential profile through the diffuse layer. Reprinted with permission from [12]. Copyright (2011) American Chemical Society.

The relationship between σ^M and potential can be evaluated according to Gauss's law using a Gaussian surface in a rectangular box near the electrode surface. The box contains all the charges in the diffuse layer, and the charge is

$$q = \varepsilon\varepsilon_0 \int \mathcal{E} \cdot dS.$$

The field strength is zero at all points on the Gaussian surface except at the electrode interface; thus,

$$q = \varepsilon\varepsilon_0 \left(\frac{d\varnothing}{dx}\right)_{x=0} \int ds$$

$$\sigma^M = -\sigma^s = (8kT\varepsilon\varepsilon_0 n^0)^{1/2} \sinh\left(\frac{ze\varnothing_0}{2kT}\right)$$

or

$$\sigma^M = 11.7C^{*1/2} \sinh(19.5z\varnothing_0),$$

where the bulk concentration of the electrolyte, C^*, is expressed in mol l^{-1}. Thus, the differential capacitance can be evaluated as follows:

$$C_d = \frac{d\sigma^M}{d\varnothing_0} = \left(\frac{2z^2e^2\varepsilon\varepsilon_0 n^0}{kT}\right)^{1/2} \cosh\frac{ze\varnothing_0}{2kT}$$

or

$$C_d = 228zC^{*1/2} \cosh(19.5z\varnothing_0)$$

where C_d is expressed in μF cm^{-2}.

Figure 1.6 shows the resulting variation of differential capacitance with the potential. It is minimum at the PZC and rises faster at extreme potentials.

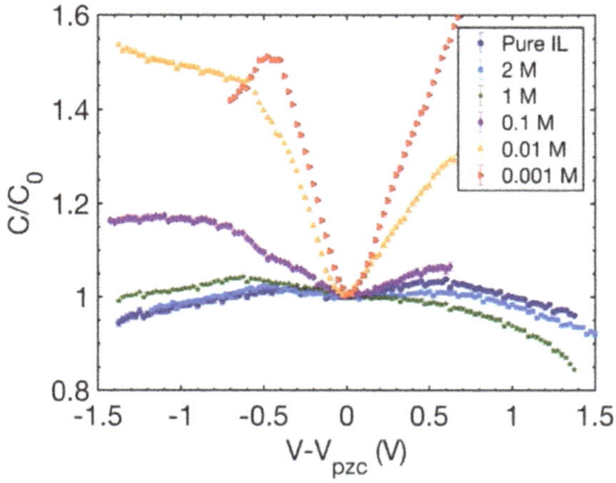

Figure 1.6. The Gouy–Chapman model predicts that the differential capacitance rises rapidly at extreme potentials. Reprinted with permission from [14]. Copyright (2017) American Chemical Society.

However, in real systems, the potential remains constant at extreme values, as also observed earlier in figure 1.4. Moreover, the actual capacitance values are much lower than those predicted by this model; therefore, this model presents an incomplete picture [13].

1.3.3 Stern's modification

The Gouy–Chapman model does not consider the finite size of ions in the electrolyte; rather, they are considered to be point charges. Therefore, the effective separation distance between the metallic electrode surface and charges in the electrolyte decreases toward zero. Real ions have a finite size, which dictates a closest approach distance equal to the ionic radius, as shown in figure 1.1.

The electrolyte side closest to the electrode is an inner layer that contains the solvent molecules and specific adsorbates (other species of ions or molecules). The inner layer is also called the Helmholtz layer or the Stern layer. The IHP is the locus of the electrical centers of the adsorbed ions. It is located at a distance x_1 from the electrode surface. The solvated ions arrange themselves along the locus of their centers in a plane (known as the OHP) and maintain a closest approach distance x_2. At low electrolyte concentrations, the thickness of the diffuse layer is large compared to x_2, while in concentrated electrolytes, the charge is compressed at x_2, which represents systems that follow the Helmholtz model. This model was suggested by Stern. Now, for $x \gg x_2$, the potential profile is given by

$$\int_{\varnothing_1}^{\varnothing} \frac{d\varnothing}{\sinh(ze\varnothing/2kT)} = -\left(\frac{8kTn^0}{\varepsilon\varepsilon_0}\right)^{1/2} \int_{x_2}^{x} dx$$

or

$$\frac{\tanh(ze\varnothing/4kT)}{\tanh(ze\varnothing_2/4kT)} = e^{-k(x-x_2)}.$$

The field strength at x_2 is given by

$$\frac{d\varnothing}{dx}_{x=x_2} = -\left(\frac{8kTn^0}{\varepsilon\varepsilon_0}\right)^{1/2} \sinh\left(\frac{ze\varnothing_2}{2kT}\right),$$

since the charge density between the electrode and the OHP is zero and the field remains the same. Thus, the potential profile across the OHP is linear for a compact layer. The potential drop across the double layer is then

$$\varnothing_0 = \varnothing_2 - \left(\frac{d\varnothing}{dx}\right)_{x=x_2} x_2.$$

The magnitude of charge is

$$\sigma^M = -\sigma^S = -\varepsilon\varepsilon_0\left(\frac{d\varnothing}{dx}\right)_{x=x_2} = (8kT\varepsilon\varepsilon_0 n^0)^{1/2} \sinh\left(\frac{ze\varnothing_2}{2kT}\right)$$

or

$$\sigma^M = (8kT\varepsilon\varepsilon_0 n^0)^{1/2} \sinh\left[\frac{ze}{2kT}\left(\varnothing_0 - \frac{\sigma^M x_2}{\varepsilon\varepsilon_0}\right)\right],$$

and

$$C_d = \frac{d\sigma^M}{d\varnothing_0} = \frac{(2z^2e^2\varepsilon\varepsilon_0 n^0/kT)^{1/2}\cosh(ze\varnothing_2/kT)}{1 + (x_2/\varepsilon\varepsilon_0)(2\varepsilon\varepsilon_0 z^2e^2n^0/kT)^{1/2}\cosh(ze\varnothing_2/2kT)}.$$

After rearranging, we obtain

$$\frac{1}{C_d} = \frac{x_2}{\varepsilon\varepsilon_0} + \frac{1}{(2\varepsilon\varepsilon_0 z^2e^2n^0/kT)^{1/2}\cosh(ze\varnothing_2/2kT)}.$$

The above equation is the sum of two capacitances in series, as depicted in a representative figure 1.7(a), as follows:

$$\frac{1}{C_d} = \frac{1}{C_H} + \frac{1}{C_D},$$

where C_H corresponds to the capacitance of the charges in the OHP, whereas C_D is the capacitance of the diffuse layer. C_H is independent of potential, while C_D varies in a V-shaped manner. This model is called the Gouy–Chapman–Stern (GCS) model. Although the model provides a valid justification for earlier models, it neglects the fact that C_H depends on potential, ion-pairing effects in the double layer, and any strong nonspecific interactions of the ions with the charge on the surface of the electrode (figure 1.7(b)).

The difference in the distance of closest approach is mainly due to the fact that common anions are bigger in size, as shown in table 1.1. Cations retain solvation shells due to strong ion–solvent dipole interactions.

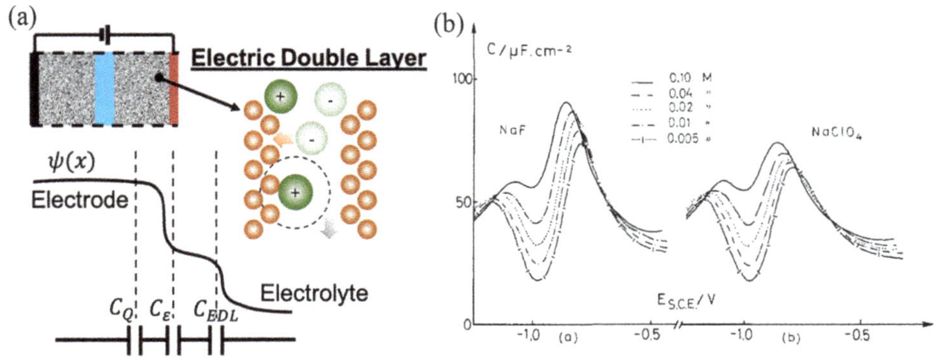

Figure 1.7. (a) Schematics of the GCS model of differential capacitance, showing the Helmholtz layer and diffuse layer capacitances in series. Reprinted with permission from [15]. Copyright (2022) American Chemical Society. (b) Concentration dependence of capacitance in (a) NaF, and (b) $NaC10_4$ solutions. Reprinted from [16], Copyright (1981), with permission from Elsevier.

Table 1.1. Ionic radii of alkali and halide ions (nm). Adopted from [30].

Ion	Radii	Ion	Radii
Li^+	0.060	F^-	0.136
Na^+	0.095	Cl^-	0.181
K^+	0.033	Br^-	0.195
Rb^+	0.148	I^-	0.216
Cs^+	0.169		
$NH_4{}^+$	0.139		
H_3O^+	0.138		

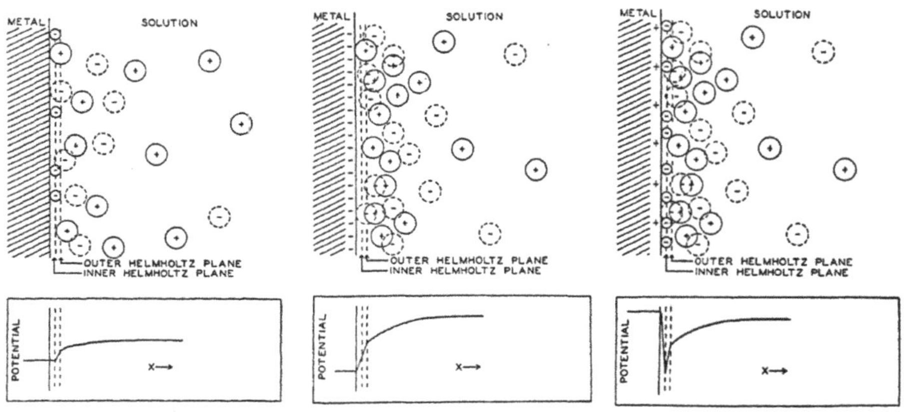

Figure 1.8. Grahame model, depicting the structure of a double layer. Reprinted with permission from [11]. Copyright (1947) American Chemical Society.

Thus, the Grahame model (figure 1.8) consists of three distinct regions: the inner and outer Helmholtz layers and a diffuse layer of ion distribution. Either of the Helmholtz layers dominates, depending on the degree of polarization (i.e. highly positive or highly negative charge density) and the population of anions or cations in relation to the polarization. It is important to note that the anion distances of closest approach are generally smaller than the hydrated cations [17]. It can be observed that the capacitance at positively charged electrode surfaces is generally twice that at a negatively charged surface. However, this also depends on the metal, the ions of the electrolyte, and the solvent [18].

1.4 Double layers in solids

The effects mentioned in earlier sections are more evident in solid electrodes (e.g. carbon or platinum). Impurities from the electrolyte can be adsorbed at the electrode surface and thus may alter the interfacial properties. Solid electrodes also have defects such as dislocation lines, varying grain sizes in polycrystalline materials, etc. There are various methods to determine whether electrodes have surface excesses of electroactive species, including cyclic voltammetry, chronocoulometry, polarography, and thin-layer methods, in addition to spectroscopic and microscopic methods.

1.5 Single crystal electrode surfaces

In polycrystalline materials, the grains consist of different crystal faces and edges, where different crystal orientations (faces and edges) come into contact with the electrolyte due to different work functions. The metals used as electrodes, such as Pt, Pd, Ag, Ni, and Cu, have a face-centered cubic (FCC) crystal structure. Low-index crystal faces such as the (100), (110), and (111) planes are the surfaces most frequently used for electrodes because of their stability [16].

Planes with high Miller indices have atomically smooth terraces, exposed edges, and kinks. Highly oriented pyrolytic graphite (HOPG) is a widely used carbon electrode. In the hexagonal close-packed (HCP) structure, carbon atoms are bonded to each other in the basal plane, which is similar to the FCC (111) plane. The properties (e.g. work function and PZC) of single-crystal electrodes strongly depend on the orientation of the plane that interfaces with the electrolyte. For example, silver electrodes present Ag (111) (−0.69 V vs. SCE) and Ag (110) (−0.98 V vs. SCE) planes. In polycrystalline electrodes, many crystal faces interacting with the electrolyte may carry different numbers of charges. In Ag, (111) sites carry negative charges, while (110) sites carry positive charges. Moreover, the crystal face also tailors the catalytic and adsorptive properties of the electrode solid surface [19].

1.6 The solid metal–solution interface

Capacitive measurements (e.g. capacitance curves for Ag(100) at different concentrations of electrolytes such as KPF_6 and NaF) can also confirm the nature of the solid/electrolyte interface, the PZC, and the differences in behavior for different crystal faces with the same electrolyte. The detection of a capacitance minimum at the PZC suggests that the PZC depends on the crystal face that interacts with the electrolyte. Surface reconstructions further complicate the interaction during potential scans [20].

1.7 Rate of specific adsorption

In specific adsorption, due to the strong interaction between the adsorbate and the electrode material, a layer forms on the electrode surface, and nonspecific adsorption takes place. In such adsorption, long-range electrostatic forces are thought to perturb the distribution of ions in the electrolyte near the electrode surface. Thus, nonspecific adsorption can affect the concentration of the electroactive species as well as the potential distribution near the electrode surface. If an electroactive species is adsorbed in a relatively higher amount than the bulk concentration originally present, then it may impact the electrochemical process, thus changing the energy of the reaction. The specific adsorption of an electroinactive species may establish a blocking layer or increase reactivity on the electrode surface of a species; whereupon the phenomenon is called electrocatalysis.

An adsorption isotherm establishes the relationship between the amount of a substance i adsorbed per unit area of the electrode, Γ_i, the activity in the bulk solution, a_i^b, and the charge or potential of the electrode, E or q^M. Assuming the electrochemical potentials of the bulk and adsorbed species i at equilibrium are equal, then

$$\mu_i^A = \mu_i^b,$$

where A and b refer to the adsorbed i and bulk i, respectively. Thus,

$$\mu_i^{0,A} + \text{RT}\ln a_i^A = \mu_i^{0,b} + \text{RT}\ln a_i^b,$$

where μ_i^0 is the standard electrochemical potential and ΔG_i^0 is the standard free energy of adsorption as a function of the electrode potential, given by

$$\Delta G_i^0 = \mu_i^{0,A} - \mu_i^{0,b}.$$

Thus,

$$a_i^A = a_i^b e^{-\Delta G_i^0/RT} = \beta_i a_i^b$$

where $\beta_i = \exp\left(\frac{-\Delta G_i^0}{RT}\right)$.

This equation represents the general adsorption isotherm. The rate at which the adsorbed layer Γ_i approaches its equilibrium value is the same as the rate of mass transfer to the electrode. Thus, according to Fick's second law of diffusion, the amount of material adsorbed at time t is [21]:

$$\Gamma_i(t) = \int_0^t D_i\left[\frac{\partial C_i(x,t)}{\partial x}\right]_{x=0} dt$$

For a stationary plane electrode, the solution of the above equation is

$$\frac{C_i(x,t)}{C_i^*} = 1 - \exp\left(\frac{x}{b_i} + \frac{D_i t}{b_i^2}\right)\text{erfc}\left[\frac{x}{2(D_i t)^{1/2}} + \frac{(D_i t)^{1/2}}{b_i}\right]$$

$$\frac{\Gamma_i(t)}{\Gamma_i} = 1 - \exp\left(\frac{D_i t}{b_i^2}\right)\text{erfc}\left[\frac{(D_i t)^{1/2}}{b_i}\right].$$

The linearized isotherm is independent of the bulk concentration.

1.8 Effect of the double layer on electrode reaction rates

The rate constants of a totally irreversible one-step, one-electron reaction are defined by the following exponential relation:

$$k_t^0 = k^0 \exp\left[\frac{-(\alpha - z)F\phi_2}{RT}\right],$$

and the exchange current, $i_{0,t}$, is defined as

$$i_{0,\,t} = i_0 \exp\left[\frac{-(\alpha - z)F\phi_2}{RT}\right],$$

where α is the transfer coefficient. Here, ϕ_2 is the potential at the OHP ($x{=}x_2$), which is not equal to the potential in the solution, ϕ^s, due to the potential drop in the diffuse layer. Thus, the potential differences affect the reaction kinetics proposed by Frumkin, called the Frumkin effect. From the above equations, it is clear that the effect of the double layer on kinetics is that the quantities k^0 and i_0 are functions of potential, due to the variation of ϕ_2 with $E - E_z$ [22].

1.9 Double layers in nonaqueous and aqueous electrolytes

In electrochemical capacitors, the solvent of the electrolyte solution constitutes the dielectric of the double layer, which, as we discussed, consists of one real, electrically conducting plate and a second virtual plate at the electrode/solution interface. The charges in the double layer are distributed across the interfacial region, a compact layer measuring 0.5–0.6 nm, which corresponds to the diameters of the solvent molecules and ions in it. Depending on the ionic concentration, a wider thermally distributed region extends up to 1–100 nm. Thus, two interfacial double layers in a device constitute its own capacitance, while in a double-layer supercapacitor, the dielectric properties of these two capacitors at the molecular level determine the specific capacitance. Intrinsically polar molecules, such as H_2O, HCl, HF, and CH_3CN, exhibit a larger degree of polarization per unit volume. H-bonded solvents offer high dielectric constant values but have the disadvantage of lower decomposition voltages of 1.23–1.5 V when used as dielectric solvents in double-layer capacitors. The use of aprotic nonaqueous solvents in electrolytes (e.g. tetraalkylammonium salts) leads to higher operating potential windows per cell. Since these solvents do not have H or other electrolytic atoms, they offer higher decomposition voltages of 3.4–4.0 V. Examples of such solvents include dimethoxyethane, acetonitrile, benzonitrile, propylene carbonate, N,N-dimethylformamide, and tetrahydrofuran. Dielectric materials that have strong intermolecular electrostatic dipole–dipole interactions between molecules such as CH_3NO_2, $(CH_3)_2CO$, and H-bonded molecules (such as liquid HF, H_2O, HCOOH, $CH_3{\cdot}COOH$, $H{-}CO{\cdot}NH_2$, esters, propylene carbonate, etc.) provide polarization between small assemblies of molecules, leading to much higher polarization per unit volume. High dielectric constants in the range of 50–113 are found in $CH_3{\cdot}OH$, H_2O, liquid HCN, liquid HF, and HCOOH.

Theoretically, the larger decomposition potential window of many nonaqueous electrolytes leads to energy densities 12–16 times larger than those of aqueous electrolytes. The differences between nonaqueous and aqueous electrolytes lie in the dielectric constants of the solvent, the dipole moments of the solvent molecules, the electron pair donor strength of various solvents and anions, and the molecular shape and size of the solvent molecules. Differences in solvation energy geometry lead to different patterns of anion adsorption. In nonaqueous solvents, the absence of H bonding and different bulk intermolecular structures compared to those of aqueous solvents affect anion adsorption. The preferred aprotic solvents for electrochemical capacitors are those without electrochemically active H atoms, such as $-NH_2$, $-OH$, and $-COOH$. These have functionalities in the molecular structure that avoid the discharge of H_2. Propylene carbonate is the most common nonaqueous solvent for high-voltage supercapacitors.

Aprotic solvents are also used with other solvents, such as dimethoxyethane, to improve the electrolytic activity of dissolved R_4N^+ salts. The capacitance behavior strongly depends on whether a solvent is structured or unstructured and the solvation state of ions. The state of the solvent structure depends on hydrogen bonding as well as strong dipole–dipole interactions in highly polar aprotic solvents. The dielectric constant and viscosity mainly relate to the extent of the solvent structure.

In double-charge layers in electrochemical capacitors, the structural aspect of solvent properties is important because the fields due to the charged electrode surface (interfacial field), the radial field, or the field gradient extending into the solution beyond the ionic periphery orient the dipole along the field. This orientation is then opposed by structural interactions such as hydrogen bonding or dipole–dipole interactions. The solvent region affected by the overlap of the electric field on the electrode surface also overlaps with the presence of ions, as shown in figure 1.9. Table 1.2 provides examples of the classification of structured, semistructured, and unstructured solvents.

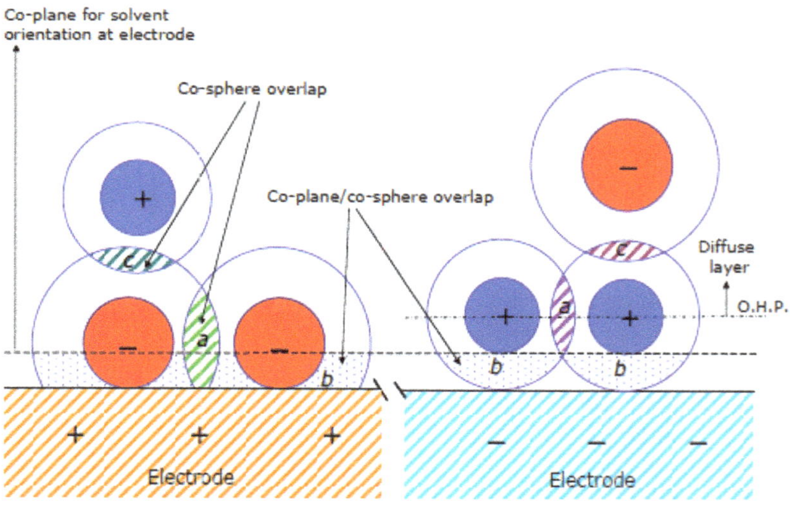

Figure 1.9. Overlap of solvent regions in the double layer at an electrode surface due to the field in the double layer and the field due to ions in the Helmholtz layer. Reprinted from [23]. Copyright (2007), with permission from Elsevier.

Table 1.2. Classification of solvents (adopted from [30]).

Structured	Intermediate	Unstructured
Water	Nitromethane	Hexane
Methanol and other alcohols	Acetonitrile	Benzene
Glycols	Propylene and ethylene carbonate	Oils
Hydrogen fluoride	N,N-dimethylformamide	2-Methyltetrahydrofuran
Hydrogen cyanide	Dimethyl sulfoxide	Ethers with large R groups
Formamide	Sulfolane	Fluorocarbons
Mono N-methyl formamide	Esters of carboxylic acids	
Formic acid	Hexamethylphosphoramide	
Acetic and other liquid carboxylic acids	Dimethoxyethane	
Amines	Dioxane and other cyclic ethers	
	Ketones, lactones, and pyridine	

The structured solvents have high dielectric constants in the range of 30–115, the semistructured solvents have dielectric constants between 10 and 20, and the unstructured solvents have low dielectric constants between three and eight.

The electrical behavior of an electrolyte solution system can be determined by its equivalent series resistance (ESR) and thus evaluating the power output capability and conductance of the electrolyte of the device. Alternatively, anion adsorption from the electrolyte and the dielectric properties of the solvent determine the specific double-layer capacitance [24]. Moreover, the conductance is also influenced by the extent of ionization or ion pairing of the solute salt. The factors that mainly influence the conductance are the concentration of free charge carriers, the ionic mobilities, and the cations/anions of any salt or acid, since each dissociated ion of the electrolyte has its own characteristic conductance. These factors further depend on the solubility of salt in the solvent and the degree of dissociation of the salt molecules into mobile ions, i.e. the extent of pairing of ions of the dissolved salt or acid. The four types of ion pairing are shown in figure 1.10.

To obtain the best power performance from a supercapacitor, its internal resistance should be minimized, and electrolyte conductance should be maximized. Hence, the electrochemically compatible electrolyte salt, acid, or alkali used should be strongly soluble in the solvent. Also, its dissociated ions in the dissolved state should exhibit maximal free mobility and minimal ion pairing. Any given type of electrolyte has a wide range of conductivities, which depend on the type of solvent and the radii and charges of the ions, which are categorized as leveling and nonleveling. Leveling solvents have strong solvation and tend to minimize ion pairing, i.e. complete dissociation. These solvents have high dielectric constants;

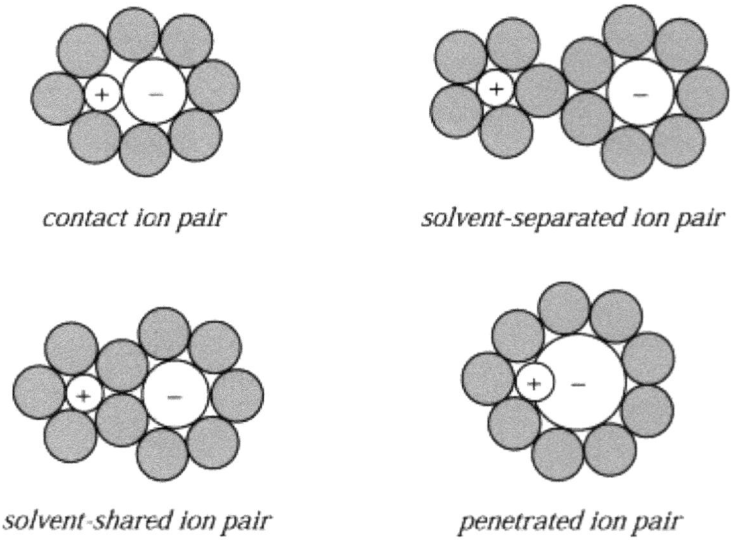

Figure 1.10. The types of ion pairs in the electrolyte solution, namely contact ion pairs, solvated ions, solvent-shared ions, and penetrated ion pairs. Reprinted with permission from [25]. Copyright (2005) American Chemical Society.

Table 1.3. Factors determining the conductance of an electrolyte (adopted from [30]).

1. Solubility of the salt and acid
2. Degree of dissociation or extent of cation and anion pairing in solution
3. Dielectric constant of the bulk solvent
4. Electron pair donor strength of the solvent molecules
5. Mobility of the free (dissociated) ions
6. Viscosity of the solvent
7. Solvation of free ions and the radii of the solvated ions
8. Temperature coefficients of viscosity and ion-pairing equilibria
9. Dielectric relaxation time of the solvent

examples include hydrogen-bonded structures with a large dipole moment. Table 1.3 summarizes the factors affecting the conductance of electrolyte solutions [26].

1.9.1 Aqueous media

For carbon-based supercapacitors, sulfuric acid or potassium hydroxide (KOH) in an aqueous medium provides a potential window of 1.23 V upon decomposition. Salts are preferred over strong acids due to the corrosive activity of acids. Due to the presence of OH^- anions, hydroxide electrolytes are water-soluble and have good conductivity [27]. The conductances of both acidic and alkaline electrolytes in aqueous media are high due to a proton-hopping mechanism [28].

1.9.2 Nonaqueous media

The larger decomposition limits of nonaqueous media provide higher potential windows; thus, they offer an added advantage over aqueous media. Tetraalkylammonium salts have good solubility in nonaqueous solvents and offer good conductivity [29]. These salts do not deposit alkali metal on the cathode of a capacitor. However, these electrolytes are costly and must remain pure and dry to avoid the formation of $2H_2 + O_2$ upon charge and recombination, which leads to self-discharge. At the negative electrode, these electrolytes can decompose on overcharge. With a suitable choice of solvent and R_4N^+ electrolytes, the operating potential window can reach as much as 4.0 V [30].

References

[1] Attard P 2001 Recent advances in the electric double layer in colloid science *Curr. Opin. Colloid Interface Sci.* **6** 366

[2] Ivanov V D 2024 The Helmholtz model *J. Solid State Electrochem.* **28** 2487

[3] Elliott J A W 2020 Gibbsian Surface Thermodynamics *J. Phys. Chem.* B **124** 10859

[4] Hansen W N 1983 The emersed double layer *J. Electroanal. Chem. Interfacial Electrochem.* **150** 133–40

[5] Mostany J, Climent V, Herrero E and Feliu J M 2010 Surface excesses at very low concentrations from extrapolation of thermodynamic data: A way to explore beyond practical limits from reliable experimental data *J. Electroanal. Chem.* **649** 119

[6] Donohue M D and Aranovich G L 1998 Classification of Gibbs adsorption isotherms *Adv. Colloid Interface Sci.* **76-77** 137

[7] Jović V D and Jović B M 2003 EIS and differential capacitance measurements onto single crystal faces in different solutions: Part I: Ag(111) in 0.01 M NaCl *J. Electroanal. Chem.* **541** 1

[8] Shibata M S, Morimoto Y, Zenyuk I V and Weber A Z 2024 Parameter-Fitting-Free Continuum Modeling of Electric Double Layer in Aqueous Electrolyte *J. Chem. Theory Comput.* **20** 6184–96

[9] Valincius G 1999 Electrocapillary equations of solid electrodes *J. Electroanal. Chem.* **478** 40

[10] Lockett V, Horne M, Sedev R, Rodopoulos T and Ralston J 2010 Differential capacitance of the double layer at the electrode/ionic liquids interface *Phys. Chem. Chem. Phys.* **12** 12499

[11] Grahame D C 1947 The Electrical Double Layer and the Theory of Electrocapillarity *Chem. Rev.* **41** 441–501

[12] Wang H and Pilon L 2011 Accurate Simulations of Electric Double Layer Capacitance of Ultramicroelectrodes *J. Phys. Chem.* C **115** 16711–9

[13] Allagui A, Benaoum H and Olendski O 2021 On the Gouy–Chapman–Stern model of the electrical double-layer structure with a generalized Boltzmann factor *Phys. A Stat. Mech. Appl.* **582** 126252

[14] Jitvisate M and Seddon J R T 2017 Direct Measurement of the Differential Capacitance of Solvent-Free and Dilute Ionic Liquids *J. Phys. Chem. Lett.* **9** 126–131

[15] Wu J 2022 Understanding the Electric Double-Layer Structure, Capacitance, and Charging Dynamics *Chem. Rev.* **122** 10821–59

[16] Valette G 1981 Double layer on silver single-crystal electrodes in contact with electrolytes having anions which present a slight specific adsorption: Part I. The (110) face *J. Electroanal. Chem. Interfacial Electrochem.* **122** 285

[17] Bockris J O, Devanathan M A V and Müller K 1965 On the structure of charged interfaces *Electrochemistry* **541** 832

[18] Hills G J and Hsieh S E 1974 Surface excess entropies and volumes of the mercury-electrolyte interface *J. Electroanal. Chem. Interfacial Electrochem.* **54** 289–98

[19] Guerrini E, Consonni V and Trasatti S 2005 Surface and electrocatalytic properties of well-defined and vicinal RuO_2 single crystal faces *J. Solid State Electrochem.* **9** 320

[20] Doubova L M, Hamelin A, Stoicoviciu L and Trasatti S 1992 Temperature dependence of double-layer parameters of the silver (100) face: Comparison with other silver faces *J. Electroanal. Chem.* **325** 197

[21] Pal B, Yang S, Ramesh S, Thangadurai V and Jose R 2019 Electrolyte selection for supercapacitive devices: a critical review *Nanoscale Adv.* **1** 3807

[22] Trachtenberg I 1957 Adsorption Kinetics and Electrode Processes *PhD* Louisiana State University and Agricultural & Mechanical College https://doi.org/10.31390/gradschool_disstheses.206

[23] Krishnan C V, Garnett M and Chu B 2007 Solute-Solvent Interactions from Admittance Measurements: Potential Induced and Water Structure-Enforced Ion-Pair Formation *Int. J. Electrochem. Sci.* **2** 958–72

[24] Ashraf H, Cao B, Kalhor P and Yu Z W 2021 Identification and properties of ion-pairs in the aqueous solutions of LiI and NaI by FTIR and quantum chemical calculations *J. Mol. Liq.* **322** 1

[25] Macchioni A 2005 Ion Pairing in Transition-Metal Organometallic Chemistry *Chem. Rev.* **105** 2039–74

[26] Mendhe A and Panda H S 2023 A review on electrolytes for supercapacitor device *Discov. Mater.* **3** 29

[27] Iqbal M Z, Zakar S and Haider S S 2020 Role of aqueous electrolytes on the performance of electrochemical energy storage device *J. Electroanal. Chem.* **858** 113793

[28] Gileadi E and Kirowa-Eisner E 2006 Electrolytic conductivity—the hopping mechanism of the proton and beyond *Electrochim. Acta* **51** 6003

[29] Prasad M R R and Sudhakarbabu K 2014 Thermal decomposition of tetraethyl ammonium tetrafluoroborate *J. Therm. Anal. Calorim.* **115** 1901

[30] Conway B E 1999 *Electrochemical Supercapacitors* (New York: Springer)

IOP Publishing

Chapter 2

Concepts of electric capacitors and electrolytic capacitors

2.1 Capacitance, electric field, and potential

Capacitor technology was invented in 1745 with the discovery of the Leyden jar. In 1896, the first electrolytic capacitor was constructed using aluminum leaf with alumina as the dielectric. Thereafter, a series of capacitors were developed; for example, paper capacitors were made by placing waxed paper between metal foils and rolling it into a cylindrical column. Later, mica-dielectric capacitors, polyethylene terephthalate–based capacitors, and plastic-dielectric capacitors were mainly used in the operation of electrical and electronic components. The unique ability of dielectric capacitors to rapidly release stored energy prompted their use in advanced applications, including medical devices, energy systems, avionics, and military systems.

Dielectric capacitors store electrostatic energy resulting from charge separation and alignment by an external electric field. The energy stored in a capacitor after multiple charging cycles can be released to generate an electric pulse of the required power. However, dielectric losses and leakage current due to conduction cause the stored energy to dissipate. Dielectric capacitors consist of a dielectric material placed between two metal electrodes in a simple configuration (figure 2.1(a)). The charge stored by dielectric capacitors is measured by their capacitance, which depends on the area (A) of contact between the electrode and the dielectric, as well as the distance (d) between the electrodes. It is given by $C = dQ/dV = \varepsilon_0\varepsilon_r A/d$, where ε_0 and ε_r denote the vacuum and relative permittivities, respectively. The volumetric energy density stored in a linear dielectric capacitor is given by $U_e = W/Ad = \int_0^{Q_{\max}} VdQ/Ad = \int_0^{D_{\max}} EdD$. Here, D is the electrical displacement ($=\varepsilon_0\varepsilon_r E$), which is related to the polarization of a dielectric material by $D = P + \varepsilon_0 E$ where, P is polarization, E is electric field and V is the potential difference. Thus, the energy density of a capacitor depends on permittivity and breakdown strength and is given

doi:10.1088/978-0-7503-5944-3ch2

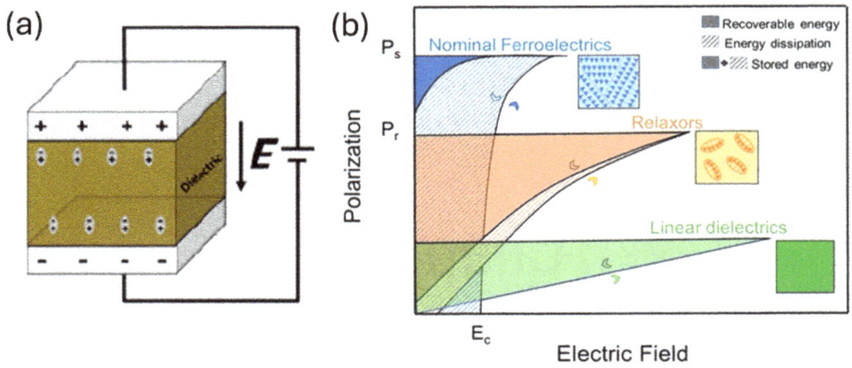

Figure 2.1. (a) Circuit diagram of a dielectric capacitor. Reproduced from [2]. CC BY 4.0. (b) Polarization–electric field (*P–E*) hysteresis loops of dielectric materials, showing recoverable stored energy density and energy loss. Reproduced from [3]. CC BY 4.0.

by $U_e = 1/2\ (\varepsilon_0\varepsilon_r E_b{}^2)$, where E_b is the breakdown strength [1]. Thus, increasing the energy density of a capacitor requires increased permittivity and polarization, along with higher breakdown strength.

Nonlinear dielectrics, namely ferroelectrics, relaxor ferroelectrics, and antiferroelectrics, exhibit energy dissipation; thus, the stored (charge) energy density (U_e) and (discharge) recoverable energy density (U_{rec}) can be evaluated from the polarization–electric field (*P–E*) hysteresis loop by integrating the effective area between the polarization axis and the charging and discharging curves, as shown in figure 2.1(b) [1].

The stored energy density and recoverable energy density can be calculated using the following relation: $U_e = \int_0^{P_{\max}} EdP$ and $U_r = \int_{P_r}^{P_{\max}} EdP$, where $P_{\max}$ and P_r are the maximum and remanent polarizations, respectively. The only challenge most of these materials face is achieving low dielectric loss and low electric/ionic conductivity. The other parameters used to evaluate energy storage performance are energy storage efficiency, dielectric breakdown, discharge time, fatigue endurance, and thermal stability [4].

2.1.1 Energy storage efficiency

During the discharge process, some of the stored energy is dissipated (U_{loss}), which can be quantified by calculating the area of the *P–E* hysteresis loop. Under high electric fields, an increase in the conduction of the material induces dielectric/polarization loss, which reduces the stored energy or the energy recovered during discharge. Furthermore, the loss of energy generates heat in the capacitor, reducing the thermal stability and lifetime of the capacitor. The energy storage efficiency (η) of a capacitor is given by $\eta = U_r/U_e = U_r/(U_r + U_{\text{loss}})$. Thus, to increase the energy storage efficiency of capacitors, their dielectric materials should have low dielectric/hysteresis loss, delayed polarization saturation, and low remnant polarization [5].

2.1.2 Dielectric breakdown field

The dielectric breakdown field (E_b) is the maximum electric field above which the resistance of the dielectric reduces significantly; thus, the value of E_b greatly affects the operation of a dielectric capacitor. The dielectric breakdown field is not only influenced by the bandgap and permittivity of the dielectric capacitor but also by its grain size, phase structure, sample thickness, porosity, and electrode configuration. When a dielectric is exposed to a swept voltage, a current flows through it due to the generation of electrons, which increases rapidly after reaching a saturation point. An increase in the current over the short period of 10^{-8} s creates an avalanche and thus breakdown [6].

2.1.3 Discharge time

The discharge time is the time taken for the energy discharged through the load to reach 90% of its final value, where the final value is measured over an infinitely long time. The discharge energy is calculated as a function of time using the time-dependent voltage and transient current and is given by $U_{\mathrm{rec}}\,(t) = 1/v \int_0^t V(t)I(t)dt = \frac{\tau V_0^2}{2vR_L}\left[1 - e^{\frac{2t}{\tau}}\right]$, where the voltage at time t is $V(t) = V_0 e^{-t/\tau}$, V_o is the maximum applied voltage, $I(t)$ is the current at time t, $\tau = R_L C$ is the theoretical relaxation time, R_L is the load resistance, C is the capacitance, and v is the volume of capacitor material. At $t \rightarrow \propto$, $U_{\mathrm{rec}} = \frac{\tau V_0^2}{2vR_L}$; thus, $U_{\mathrm{rec}}(t) = U_{\mathrm{rec}}(\propto)\,[1 - e^{\frac{2t}{\tau}}]$. The time in which a capacitor can discharge energy not only depends on the permittivity and thickness of the dielectric material but also on working conditions such as load resistance and applied voltage [7].

2.2 Polarization mechanisms

Various dielectric materials have been tested for energy storage capacitor applications. These are categorized as linear dielectrics, paraelectrics, ferroelectrics, relaxor ferroelectrics, and antiferroelectrics. Linear dielectrics show almost zero permanent dipoles; therefore, they exhibit constant permittivity (ε_r–E) and a linear change in polarization under an applied field (P–E). Parametric materials, on the other hand, contain permanent dipoles without ferroelectric domains; they therefore have nonlinear responses of ε_r–E and P–E characteristics. Ferroelectric materials exhibit large hysteresis loops due to the presence of a net dipole moment in the absence of an electric field. The small domains in relaxor ferroelectrics have reduced coupling, resulting in limited spontaneous polarization and thus smaller hysteresis loops. Antiferroelectrics possess opposite polarizations in neighboring domains. Under a high electric field, these show double hysteresis due to reversible phase transitions.

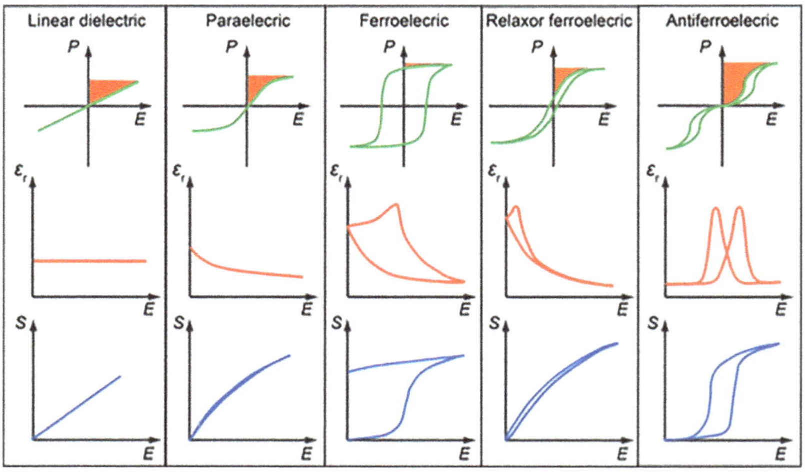

Figure 2.2. Schematics showing the behaviors of dielectric materials: (a) paraelectric, (b) anti-ferroelectric, (c) ferroelectric, and (d) relaxor ferroelectric ceramics. Reproduced from [8].

Figure 2.2 schematically illustrates the behaviors of all the described dielectric materials.

To obtain high energy storage performance from a capacitor, it is necessary to use a dielectric material with versatile properties such as high permittivity (polarizability), low dielectric loss, low ionic/electronic conductivity, and a large breakdown field. The dielectric strength is affected by many factors such as temperature, humidity, and frequency. The dielectric constant of any dielectric material can be calculated from the ratio of the cell voltage in a vacuum to the voltage with the dielectric present, $\varepsilon_r = V_0/V$. Since V is always greater than V_0, the dielectric constant is always greater than or equal to one. Table 2.1 lists the dielectric constants of various dielectric materials [9, 10].

2.3 Dielectric materials

Much effort has been invested in optimizing the properties of dielectric materials for high energy storage capabilities. These dielectrics are commonly based on ceramics, polymers, and glasses. Polymer dielectrics exemplify the shortcomings that led to the development of new dielectrics. These dielectrics have a strong breakdown field, low dielectric loss, and mechanical flexibility, but their weak thermal stability and low permittivity are limitations in various practical applications.

Popular polymer dielectrics for capacitors include polyvinylidene fluoride (PVDF) and its copolymers, biaxially oriented polypropylene (BOPP), polyethylene terephthalate, polycarbonate, and polyimide. The blending of PVDF with poly (methyl methacrylate) (PMMA) enhances energy storage performance, as shown in figure 2.3 [4]. Ceramic dielectrics, on the other hand, provide high permittivity with good thermal stability but suffer from a low breakdown field. Thus, an alternative approach for improving the energy storage capability of capacitors is to blend

Table 2.1. Dielectric constants of selected capacitor materials [11].

Dielectric	Dielectric constant
Air	1.00054
Paper	2.0 to 6.0
Plastics	2.1 to 6.0
Low-density polyethylene (LDPE)/high-density polyethylene (HDPE)	2.3
Mineral oil	2.2 to 2.3
Silicon oil	2.7 to 2.8
Quartz	3.8 to 4.4
Glass	4.8 to 8.0
Mica	7
Silicone	12
Germanium	16
Tantalum pentoxide	26
Ethanol	25
Ethylene glycol	3.7
Dipropyl ketone	12.6
Glycol	37
Glycerol	42.5
Water (20 °C)	80.4
Acetic acid	6.2
Butyric acid	3
Lactic acid	22
Cresol	5 to 11.5
Propanoic acid	3.1
Ceramics	12 to 400,000
Aluminum oxide	4.5 to 115
Titanium oxide	14 to 110
Porcelain	5.1 to 8.0
Zirconia	7.1 to 10.5
Titanates (Ba, Sr, Ca, Mg, Pb)	15 to 12,000

materials, such as ceramics and polymer-based composites, to make dielectric composite materials. Both polymer and ceramic-based dielectrics are classified according to different characteristics [12, 13].

2.3.1 Ceramic-based dielectrics

Ceramic dielectric materials are insulators such as titanates, porcelains, and metal oxides. Linear dielectric ceramics have much smaller breakdown fields than those of

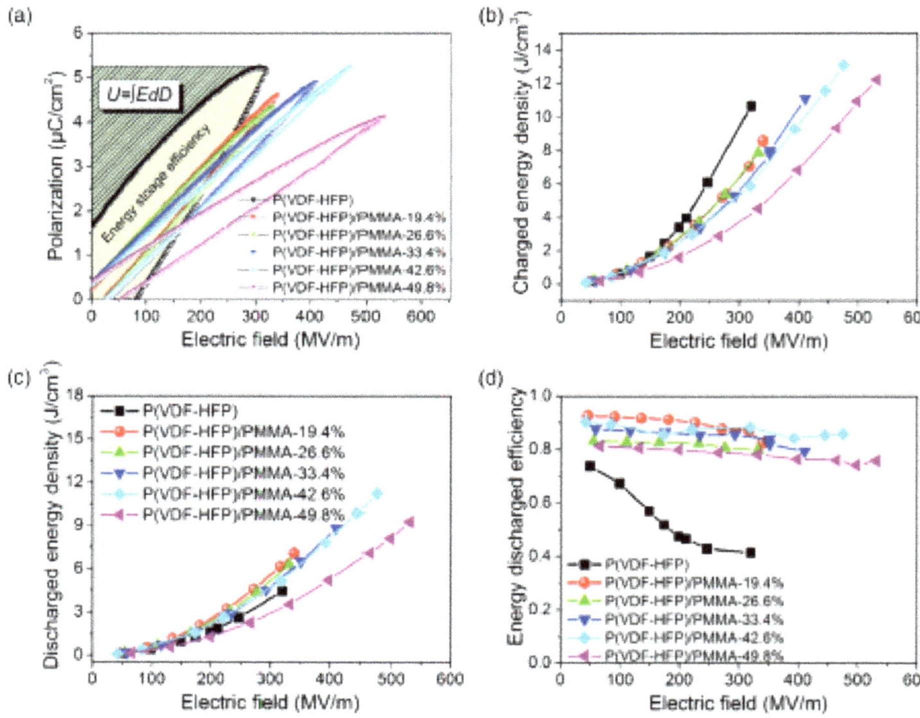

Figure 2.3. Polarization characterizations of PVDF/PMMA blends. Reproduced from [13]. CC BY 4.0.

polymers but have higher permittivity. Nonlinear dielectric ceramics offer good thermal stability for high-temperature applications (250 °C). Most bulk ceramics present low energy densities ($<10\ \mathrm{J\,cm^{-3}}$) due to their low breakdown fields ($<1\ \mathrm{MV\,cm^{-1}}$). However, dielectric ceramic films demonstrate excellent energy storage performance ($>100\ \mathrm{J\,cm^{-3}}$) and can tolerate a high applied electric field of $>1\ \mathrm{MV\,cm^{-1}}$. An interesting inverse correlation is found between the dielectric strength and the thickness of a sample. At a critical thickness, the dielectric strength becomes independent of sample dimensions. Therefore, dielectric films have higher dielectric strength and energy storage density than the bulk electric counterparts. Ceramic film capacitors have attracted wide interest in the fields of microelectronics, power devices, and communication systems. The thickness- and volume-limited total energy of film capacitors allows them to combine optimum properties such as high polarization, low hysteretic loss, high dielectric constants, low dielectric loss, large breakdown field strengths, thermal stability, and high fatigue endurance, thus maximizing energy storage.

The types of ceramic films include linear dielectric ceramic films, paraelectric ceramic films, ferroelectric ceramic films, relaxor ferroelectric films, and antiferroelectric ceramic films. Table 2.2 lists the different approaches used to enhance the energy storage performance of dielectric ceramic films [15].

Table 2.2. Various modifications made to ceramic films for enhanced energy storage. Adopted from [16].

Physical modifications	Chemical modifications	Microstructural modifications
Film thickness control	Composition optimization	Phase structure tuning
Compositional gradient sequence of layers	Elemental doping and chemical substitution	Crystallinity, grain size, defects, and impurities
Substrate material and its orientation	Low-melting-point glass additives	Crystallographic orientation
Top and bottom electrodes	Alloying/composite/solid solution	Domain engineering
Buffer layer, seed layer, insert layer, capping layer	Humidity/moisture control	Interface engineering—epitaxial or misfit strain, charge transport, and residual stress/strain

2.3.2 Polymer- and glass-based dielectrics

Polymer dielectrics add the benefits of mechanical flexibility, easier fabrication steps, low processing cost, and high breakdown fields (>6 MV cm^{-1}). However, their low dielectric permittivity, poor operational temperatures (<100 °C), and poor thermal stability are limiting factors for their application in energy storage. Dielectric polymers such as PVDF and its copolymers, BOPP, polyethylene terephthalate, polycarbonate, polyimide, etc. are also being used along with high-permittivity ceramics dispersed in a polymer matrix with a high breakdown field strength. Other high-permittivity dopants include graphene, carbon black, and carbon nanotubes, but agglomeration causes major issues for their dispersion in the matrix. Glass-based dielectrics possess greater permittivity and tolerate much larger breakdown fields (>10 MV cm^{-1}) compared to the polymer dielectrics [17].

2.4 Types of capacitors

Various capacitors are designed using different dielectric materials to suit specific applications. Capacitors are categorized as ferroelectric, polar polymer dielectric, linear, nonlinear, fixed, variable, power, high-voltage, and interference-suppression capacitors [18].

2.4.1 Linear and nonlinear capacitors

In linear capacitors, the capacitance does not change with the operating voltage and frequency. These are the most widely used capacitors (Class I). Class I materials are natural dielectrics, such as glass and mica. However, if the capacitance changes when the operating voltage and frequency are altered, the capacitor is classified as nonlinear (Class II) and is used for specific niche applications. Examples of Class II materials include ferroelectric ceramics and polymer electrolytes with a high dielectric constant. These are employed to obtain high capacitances.

2.4.2 Ferroelectric capacitors

Ferroelectric capacitors use ceramic dielectrics due to their high dielectric constants (1000 < ε_r <30 000). These are nonlinear capacitors (Class II). The capacitance of a ferroelectric ceramic capacitor changes with the operating voltage and frequency; thus, the capacitance shows a strong dependence on these parameters. These are called nonlinear capacitors.

2.4.3 Polar polymer dielectric capacitors

Polar polymer dielectric capacitors are fabricated using physical vapor deposition to coat polycarbonate foil with thin aluminum layers that act as electrodes. Melted metal is deposited using an airbrush method onto a cylindrical or flattened roll to provide electrical contacts for the connections with the terminals. These are mostly used in miniaturized circuits in modern electronics.

2.4.4 Fixed and variable capacitors

In fixed capacitors, the capacitance cannot be adjusted. Fixed capacitors use linear dielectrics, and their capacitance should not change with variations in charge quantity, cell voltage, or frequency. A variable capacitor is constructed to allow the capacitance to be adjusted manually. A variable capacitor with an inductor allows one to attain a desired resonance at a radio station frequency in radio applications. These capacitors have low capacitances in the range of 10–500 pF and are used for frequency filtering. The dielectric materials used in these capacitors are air, plastic foil, inert gases, etc.

2.4.5 Power capacitors

Compared to fixed capacitors, power capacitors are large, with a typical weight of 10–50 kg. The dielectric materials used in power capacitors are paper, polypropylene, synthetic oils, etc. The electrodes can be aluminum foils or thin metals deposited on the dielectrics. However, overheating may cause long-term issues with reliability and longevity. These capacitors are used in electrical engines as starter motor capacitors.

2.4.6 High-voltage capacitors

Capacitors used for high-frequency applications are mostly ceramic capacitors. A series of connections of several elementary capacitors with matching capacitances can reduce the risk caused by high voltages in AC circuits. Tube-shaped capacitors can achieve voltage ratings of up to 6 kV.

2.4.7 Interference-suppression capacitors

These capacitors are separated into classes depending on their use: Class X and Class Y. Lower safety requirements can be met by Class X capacitors, while Class Y capacitors are used to meet higher safety requirements. Both types are available in

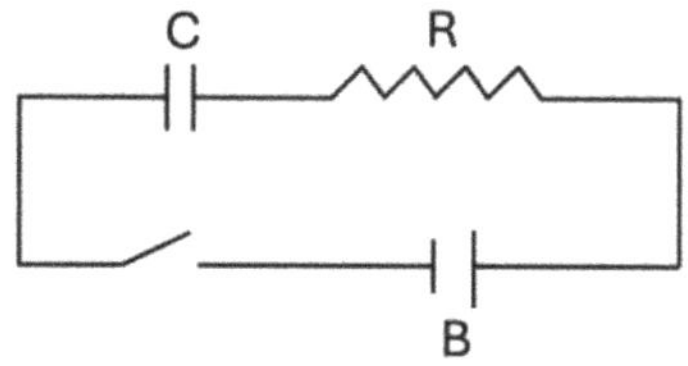

Figure 2.4. A series circuit consisting of a capacitor and a resistor connected to a battery.

materials such as ceramics (for high frequencies) and polymers (for low frequencies) and are used in radio-frequency and automotive applications [19].

2.5 Charging and discharging of capacitors

Capacitors are charged by an external power supply (e.g. a battery) in a closed-loop circuit. During the charging process, one capacitor plate becomes positively charged after losing electrons; simultaneously, the other plate accumulates an equal number of electrons flowing from the negative terminal of the battery and becomes negatively charged. The charging process stops when the potential between the plates equals the potential difference between the terminals of the power supply. A resistor, *R*, placed between a capacitor, *C*, and a power supply increases the charging time compared to that of a direct connection. The *RC* time constant is an important quantity used for measuring capacitor charging, giving charge and relaxation times. The *RC* constant can be evaluated from the circuit given in figure 2.4 using Kirchhoff's law, $V - IR - Q/C = 0$. Here, $I = dq/dt$ is a first-order differential equation. The charge q can now be calculated using $q = CV(1 - e^{t/RC})$; $q = 0$ at $t = 0$, and $q = CV$ as time approaches infinity. Therefore, the potential across the capacitor plates (V') for an applied potential of V is $V' = V(1 - e^{-t/RC})$, where RC is known as the capacitive time constant $\tau = RC$ [20].

After a capacitor is charged to the potential of the power source, a discharge can be initiated by completing the circuit with a resistor, *R*. In this circuit, the sum of potentials is $q/c + (dq/dt)R = 0$ and the charge q is given by $q = q_0 e^{-t/RC}$, where $q_0 = VC$ and the discharge current is given by $I = -(q_0/RC)e^{-t/RC}$. The charge–discharge curve is shown in figure 2.5.

2.6 Energy storage mechanism and electrical circuits

In a capacitor, an applied potential accumulates charge on the metallic plates, and as the electric field develops, it becomes difficult to accumulate charge further. An external power source provides electric potential energy, *E*, which is stored within the dielectric material. During the discharge of the capacitor, the stored energy is retrieved. Thus, the work done in the charging process is given by $dW = Vdq = (q/C)dq$; therefore, the net work done (i.e. the stored potential energy in the capacitor) is given by $E = q^2/2C = 1/2\ CV^2$. Capacitors can be arranged in series and/or parallel combinations in a single-loop circuit, as shown in figure 2.6. In parallel combinations of capacitors, all capacitors experience the same equivalent potential, V. The total charge of the parallel combination is then given by $Q = Q_1 + Q_2 + Q_3 + \cdots = (C_1 + C_2 + C_3 + \cdots)V$,

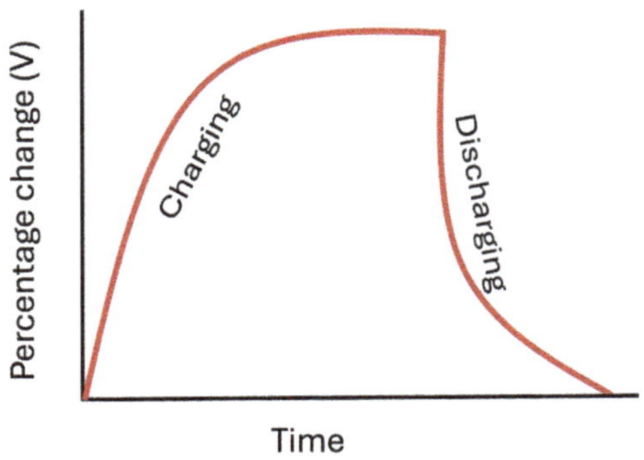

Figure 2.5. Charge–discharge curve of a capacitor versus the time.

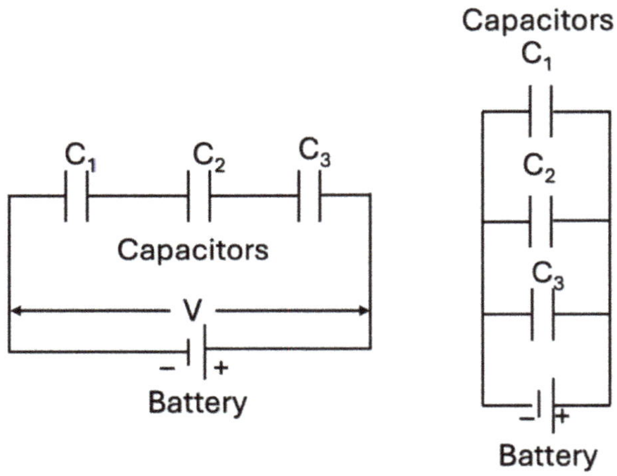

Figure 2.6. (a) Series and (b) parallel combinations of capacitors in a loop circuit.

and the equivalent capacitance is given by $C_p = Q/V = \sum_{i=1}^{n} C_i$. On the other hand, for capacitors connected to a potential difference in series, V charges each capacitor by the same amount. Thus, the total potential across the series-connected capacitors is $V' = \sum_{i=1}^{n} V_i = Q\left(\sum_{i=1}^{n} \frac{1}{C_i}\right) = \frac{Q}{C_S}$. Therefore, the equivalent capacitance of capacitors connected in series is smaller than the capacitance of any of the connected capacitors [21].

Capacitors are used along with resistors and inductors in various combinations of circuits as follows: (1) two-element circuits (inductor–capacitor (*LC*) and resistor–capacitor (*RC*) circuits) and (2) three-element circuits (resistor–inductor–capacitor circuits). We have already discussed the details of *RC* circuits in the previous section. An inductor (*L*) is similar to a capacitor in that it stores energy; however, it uses an induced magnetic field. In both *RC* and RL circuits, the growth or decay of the charge is exponential. However, *LC* circuits behave differently due to new parameters: an angular frequency, ω, and a sinusoidal oscillation period, *T*. In LC circuits, when the external power source is removed, the capacitor discharges through the inductor. As a result, the capacitor is charged to a maximum potential, and the energy stored in the inductors' magnetic field is reduced to zero. Upon discharge of the capacitor, the inductor stores the charge in a magnetic field and then periodically releases it to charge the capacitor. This is called a tank circuit. The energy of an *LC* circuit is given by $E = E_C + E_L = LI^2/2 + q^2/2C$, where E_C and E_L are the stored energies of the capacitor and the inductor, respectively. If there is no energy loss, then $dE/dt = 0$ and $I = dq/dt$; thus, $dI/dt = d^2q/dt^2$. In an ideal system, the solution of the differential equation $dE/dt = Ld^2q/dt^2 + (1/C)q = 0$ for no loss of energy gives the maximum charge Q at time $t = 0$, as $q = Q\cos(\omega t + \varphi)$, where ω is the angular frequency of oscillation and φ is the phase angle. Here, $I = -\omega Q\sin(\omega t + \varphi)$, where ωQ is the equivalent maximum current *I* passing through the circuit, and ω is $1/\sqrt{LC}$, known as the natural angular frequency of oscillation of an *LC* circuit.

A three-element circuit containing the three components of resistance, inductance, and capacitance is called an *RLC* circuit. The integration of resistance into an *LC* circuit dissipates the oscillating electromagnetic energy as thermal energy. Thus, the differential equation that considers these changes is $dE/dt = Ld^2q/dt^2 + Rdq/dt + (1/C)q = 0$. This second-order differential equation can be solved to obtain $q = Qe^{-Rt/2L}\cos(\omega' t + \varphi)$. The angular frequency is now changed due to the presence of *R* and is given by $\omega' = \sqrt{\omega^2 - \left(\frac{R}{2L}\right)^2}$. An additional external power source must provide a potential to compensate for the dissipation of thermal energy through the resistor.

2.7 Types of electrolytic capacitors

In 1886, Charles Pollak was the first to use an electrolyte as a dielectric following his investigation of the anodization of metals. In electrolytic capacitors, the dielectric is a metal oxide, such as aluminum oxide, tantalum pentoxide, or niobium oxide, which is used in combination with an aqueous or nonaqueous liquid electrolyte. Among the aqueous liquid electrolytes typically used are boric acid or sodium borate in aqueous solution. Aluminum electrolytic capacitors use a standard water-based electrolyte containing ethylene glycol and dipropyl ketone. Nonaqueous liquid electrolytes are generally composed of weak acids—primarily organic acids such as glacial acetic acid, propionic acid, lactic acid, acrylic acid, butyric acid, crotonic acid, phenol, and cresol. They may also include weak inorganic acids such as sodium perborate and trisodium phosphate. Salts derived from weak acids, such as ammonium or metal salts of organic acids, include ammonium acetate, ammonium citrate, aluminum acetate, calcium lactate, and ammonium oxalate. Electrolyte solvents are based on polyols (diethylene glycol and glycerol) or alkanolamines (monoethanolamine, diethanolamine, and triethanolamine).

There are two main types of electrolytic capacitors: aluminum and tantalum electrolytic capacitors. In aluminum electrolytic capacitors, two conducting aluminum foils are used, where one is coated with an insulating oxide layer to form the anode and the other foil is used as the cathode. Both are separated by paper soaked in a suitable electrolyte. Tantalum electrolytic capacitors can be wet or dry, depending on whether the counter electrode is sulfuric acid or a manganese dioxide film. Dry tantalum electrolytic capacitors are used in systems that require high reliability and long lifetimes, such as computers and microsystems. Compared to aluminum capacitors, dry tantalum electrolytic capacitors have a larger capacitance-to-volume ratio. Electrolytic capacitors provide many advantages over other capacitors, such as passing AC and blocking DC when used as coupling capacitors, supporting high-current, low-frequency circuits, and offering larger capacitance per unit volume. However, their drawbacks include low breakdown voltages, higher leakage currents, shorter lifetimes with poorer tolerance and temperature ranges than those of other capacitors [21].

References

[1] Zhang X, Pu Y, Gao P, Huang X, Ma J, Zhang L and Liu Z 2024 Linear dielectric ceramics for near-zero loss high-capacitance energy storage *Mater. Today Phys.* **49** 101579

[2] Fan X, Wang J, Yuan H, Zheng Z, Zhang J and Zhu K 2023 Multi-scale synergic optimization strategy for dielectric energy storage ceramics *J. Adv. Ceram.* **12** 649–80

[3] Veerapandiyan V, Benes F, Gindel T and Deluca M 2020 Strategies to Improve the Energy Storage Properties of Perovskite Lead-Free Relaxor Ferroelectrics: A Review *Materials* **13** 5742

[4] Pan H, Jiang Y and MacManus-Driscoll J L 2023 Interplay of polarization, strength, and loss in dielectric films for capacitive energy storage: Current status and future directions *J. Mater.* **9** 516

[5] Pedroli F, Flocchini A, Marrani A, Le M Q, Sanseau O, Cottinet P J and Capsal J F 2020 Boosted energy-storage efficiency by controlling conduction loss of multilayered polymeric capacitors *Mater. Des.* **192** 108712

[6] Zhang T, Lei Y, Yin J, Du J and Yu P 2019 Effects of pores on dielectric breakdown of alumina ceramics under AC electric field *Ceram. Int.* **45** 13951

[7] Pattipaka S, Lim Y, Son Y H, Bae Y M, Peddigari M and Hwang G T 2024 Ceramic-Based Dielectric Materials for Energy Storage Capacitor Applications *Materials (Basel)* **17** 1

[8] Patel S, Chauhan A and Vaish R 2014 Enhancing electrical energy storage density in anti-ferroelectric ceramics using ferroelastic domain switching *Mater. Res. Express* **1** 045502

[9] Li D, Zeng X, Li Z, Shen Z Y, Hao H, Luo W, Wang X, Song F, Wang Z and Li Y 2021 Progress and perspectives in dielectric energy storage ceramics *J. Adv. Ceram.* **10** 675

[10] Jayakrishnan A R, Anju B A, Nair S K P, Dutta S and Silva J P B 2024 Recent development of lead-free relaxor ferroelectric and antiferroelectric thin films as energy storage dielectric capacitors *J. Eur. Ceram. Soc.* **44** 4332

[11] Yu A, Chabot V and Zhang J 2013 *Electrochemical Supercapacitors for Energy Storage and Delivery: Fundamentals and Applications* (CRC Press)

[12] Serhan M, Sprowls M, Jackemeyer D, Long M, Perez I D, Maret W, Tao N and Forzani E 2019 Total iron measurement in human serum with a smartphone *AIChE Annu. Meet. Conf. Proc.* **2019** 15192–95

[13] Aklujkar P S *et al* 2025 Rationally designed high-temperature polymer dielectrics for capacitive energy storage: An experimental and computational alliance *Prog. Polym. Sci.* **161** 101931

[14] Sun J, Luo B and Li H 2022 A Review on the Conventional Capacitors, Supercapacitors, and Emerging Hybrid Ion Capacitors: Past, Present, and Future *Adv. Energy Sustain. Res.* **3** 2100191

[15] Pan H, Kursumovic A, Lin Y H, Nan C W and MacManus-Driscoll J L 2020 Dielectric films for high performance capacitive energy storage: multiscale engineering *Nanoscale* **12** 19582

[16] Palneedi H, Peddigari M, Hwang G-T, Jeong D-Y and Ryu J 2018 High-performance dielectric ceramic films for energy storage capacitors: progress and outlook *Adv. Funct. Mater.* **28** 180665

[17] Liu X J, Zheng M S, Chen G, Dang Z M and Zha J W 2022 High-temperature polyimide dielectric materials for energy storage: theory, design, preparation and properties *Energy Environ. Sci.* **15** 56

[18] Sharma P and Kumar V 2020 Current Technology of Supercapacitors: A Review *J. Electron. Mater.* **49** 3520

[19] Negroiu R, Svasta P, Pirvu C, Vasile A and Marghescu C 2017 Electrochemical impedance spectroscopy for different types of supercapacitors *Proc. Int. Spring Semin. Electron. Technol.* **20** 1–4

[20] Xia S and Chen L 2017 Theoretical and experimental investigation of optimal capacitor charging process in RC circuit *Eur. Phys. J. Plus* **132** 235

[21] Both J 2015 The modern era of aluminum electrolytic capacitors *IEEE Electr. Insul. Mag.* **31** 24

IOP Publishing

Supercapacitors
Concepts and advances
Abha Misra

Chapter 3

Concepts of electrochemical processes

3.1 Introduction and overview of electrode processes

Electrochemistry mainly deals with chemical and electrical changes due to the passage of electric current and the induction of chemical reactions to produce electrical energy. It is important to learn the basics of electrochemical processes, even if the discussion is limited to supercapacitors in the remaining chapters. Advances in electrochemistry include several different phenomena such as electrophoresis and corrosion, the development of devices like electrochromic displays, batteries, and fuel cells, technologies to electroplate metals, and the large-scale production of materials from waste. Although each of the described phenomena and developments exploits different chemical systems, the fundamentals of their electrochemical processes remain the same, as they depend on common principles, e.g. thermodynamics or reaction rates. Electrochemistry also helps in identifying many unknown species present in, for example, human blood cells and finding trace amounts of metal ions or organic species in drinking water [1].

The main components of electrochemical cells are electrodes (electronic conductors) and an electrolyte (an ionic conductor). Charge transport occurs at the interfaces between chemical phases. Electrode materials include solid metals (e.g. Au, Pt), liquid metals (amalgams, Hg), graphite in carbon electrodes, and semiconductors (indium-tin oxide, Si). Electrolytes provide a conductive path through the movement of ions; therefore, they should offer low resistance (i.e. conductive behavior). A wide variety of electrolytes are used in different systems depending on the application; among these are liquid solutions containing H^+, Na^+, and Cl^- in water or a nonaqueous solvent. In addition, fused salts, ionically conducting polymers, and solid electrolytes are also used. Electrochemical cells generally consist of two electrodes separated by an electrolyte phase in a simple geometry. However, the electrode geometry may vary for different applications. Measurement and control of the cell potential play crucial role in determining the relative energies of the carriers and also control the direction and rate of charge transfer. The reader

doi:10.1088/978-0-7503-5944-3ch3

is referred to references [1, 2] for more details about the electrochemical methods described later in this chapter.

3.2 Kinetics of electrode processes

The chemical reaction in a cell takes place between two independent electrodes. The reaction at each electrode corresponds to a half-cell reaction of the overall chemical reaction, which is the result of an interfacial potential difference at the corresponding electrode. The electrode at which a desired reaction occurs is known as the working electrode, while the other half of the cell uses a standard electrode known as the reference electrode. The reference electrode maintains a constant composition and potential. Some of the known reference electrodes are the standard hydrogen electrode (SHE) and the normal hydrogen electrode (NHE), $Pt/H_2(a = 1)/H^+$ $(a = 1$, aqueous). The saturated calomel electrode (SCE) is a common electrode that has a potential of 0.242 V vs. the NHE and consists of $Hg/HgCl_2/KCl$ (a saturated aqueous solution). Another electrode with a potential of 0.197 V vs. the NHE is the silver–silver chloride electrode, Ag/AgCl/KCl (a saturated aqueous solution). Figure 3.1 shows a schematic representation of the typical electrochemical cell used for cyclic voltammetry (CV) measurements [3].

Thus, any changes in the potential of the working electrode are measured with respect to the reference electrode, which is similar to controlling the energy of the electrons. For example, if the working electrode supplies a negative potential relative to the reference electrode, there is an increase in the energy of the electrons; thus, a reduction current flows (from the electrode to the solution). Conversely, applying a positive potential reduces the energy of the electrons, and an oxidation current flows from the solution to the electrode. The potentials at which these processes occur are called standard potentials, E^0, as shown in figure 3.2.

In a simple cell geometry, a working electrode and a reference electrode are immersed in an electrolyte, and a potential difference is applied using an external

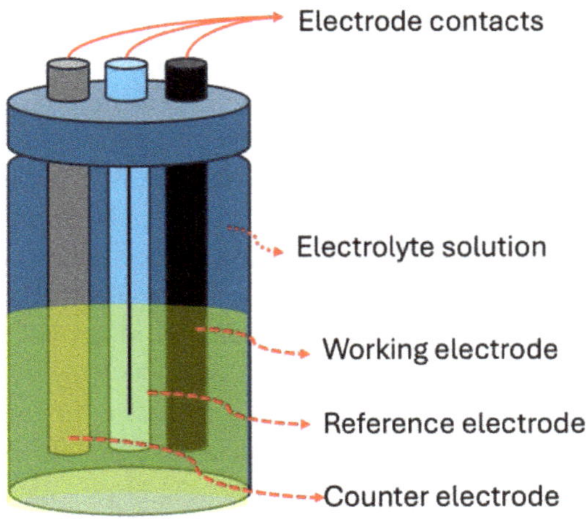

Figure 3.1. Schematic diagram showing a typical electrochemical cell.

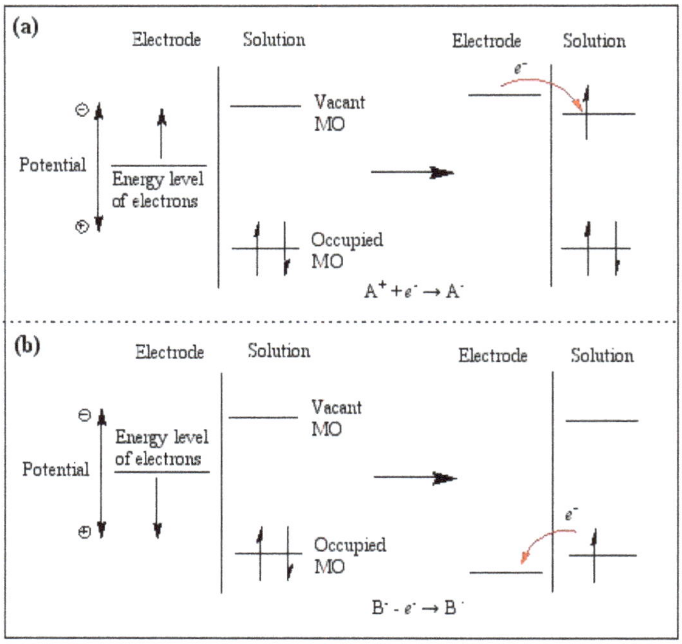

Figure 3.2. A simple representation of the oxidation and reduction processes that take place at the electrode and the electrolyte. Reproduced from [4]. CC BY 4.0.

power supply. The variation in potential, E, as the reaction occurs at the interface between the electrode and the electrolyte can produce a current in the external circuit. Thus, the qualitative results obtained from current–potential curves can be used to evaluate the nature of the reaction.

Typically, two types of charge interactions occur at the electrodes. One kind of electron transfer across the electrode–electrolyte interface results in an oxidation or reduction process. This type of interaction is known as a faradic interaction, and the electrode where this process occurs is called a charge transfer electrode. In non-faradic processes, no charge transfer reactions happen, as these are kinetically or thermodynamically unfavorable [5]. Although there is no charge transfer across the interface between the electrode and the electrolyte, external currents can still flow due to changes in potential, electrolyte composition, or the area of the electrode. Nonfaradic processes provide information related to charge transfer and the associated chemical interactions. Ideal polarized electrodes (IPEs) are electrodes where no charge transfer occurs across the electrode–electrolyte interface, irrespective of the externally applied potential. The charge at an IPE interface cannot be transferred as a result of an applied potential; thus, such an interface is similar to a capacitor [6]. At a given potential, if q^M is the excess or deficiency of electrons (charge) in a very thin layer (<0.1 A) on the surface of a metal electrode and q^S is the excess of cations or anions in the electrolyte (in the close vicinity of the electrode), then $q^S = -q^M$ at all times. The dipoles at the metal–electrolyte interface are known as an electrical double layer, and its capacitance is denoted by the double-layer capacitance, C_d, which is typically a function of potential; in this respect, it is unlike an electrostatic capacitor.

3.3 Basic potential methods

As mentioned earlier, in electrochemical systems, an electrical perturbation results in changes in the characteristics of the system. We consider the response of an IPE in an electrochemical system represented by the circuit elements R_s and C_d in series, where R_s represents the resistance of the solution and C_d a capacitor at the interface. In response to a potential step of magnitude E applied to the IPE, the behavior of the current I at time t is given by $i = \frac{E}{R_S} e^{-t/R_s C_d}$. This is the general equation for an electrostatic capacitor.

Thus, a step in potential input results in an exponentially decaying current with a time constant given by $\tau = R_S C_d$. If a constant current is used to charge an $R_S C_d$ circuit, then from the above, $E = E_R + E_C = iR_S + q/C_d$ or $E = iR_S + \frac{i}{C_d}\int_0^t dt$ or $E = i\left(R_s + \frac{t}{C_d}\right)$. Hence, the potential for a current step increases linearly over time. In response to a voltage ramp or linear potential sweep that increases linearly over time at a sweep rate of v (Vs^{-1}), the potential at time t is $E = vt$. Then, $\nu t = R_s\left(\frac{dq}{dt}\right) + q/C_d$. The solution of this equation for $q = 0$ at $t = 0$ is $i = \nu C_d\left[1 - \exp\left(-t/R_s C_d\right)\right]$. The resulting current increases from zero as the potential scan starts and attains a steady-state value, vC_d. For a triangular wave, the steady-state current changes from vC_d to $-vC_d$ during the reverse scan of the potential. Thus, in experiments, we record the current as a function of potential, which is known as linear potential sweep chronoamperometry or linear sweep voltammetry (LSV). CV is a reversal method and is a potential scan, which is same as double potential step chronoamperometry [6]. For a constant electrode area, the charging current lowers after a time (a few time constants $R_u C_d$). When the potential changes continuously in a potential sweep experiment, the charging current, i_c, and the faradic peak current, i_p, are given by [7]

$$|i_c| = AC_d\nu$$

and

$$i_p = (2.69 \times 10^5)n^{3/2}AD_o^{1/2}C_o^*\nu^{1/2}$$

Then,

$$\frac{|i_c|}{i_p} = \frac{C_d\nu^{\frac{1}{2}}(10^{-5})}{2.69n^{3/2}D_o^{1/2}C_o^*},$$

where D_o is the diffusion coefficient ($cm^2\ s^{-1}$) and C_o^* is the concentration (mol cm^{-3}). The faradic current is measured from the baseline of the charging current. While i_p varies with $v^{1/2}$ for linear diffusion, i_c varies with v. Thus, for low concentrations and high scan rates, i_c dominates, and the effect of the double layer prevails. This also shows that surface effects are dominant at high scan rates, while at small v, they are of minor importance. CV is carried out by switching the direction of the scan (at $t=\lambda$). In linear scan voltammetry, the potential at any time is given by

$$(0 < t \leqslant \lambda) \quad E = E_i - \nu t$$

$$(o < \lambda) \quad E = E_i - 2\nu\lambda + \nu t$$

The sweep reversal method known as CV is the most widely used method in electrochemistry. Upon reversal, in the *i–E* curve (cyclic voltammograms), the sign of *dE/dt* changes while its magnitude remains constant, and the charging current retains the same magnitude. In cyclic voltammograms, the faradic response is superimposed on an approximately constant charging current [8].

3.4 Electrochemical instrumentation

As shown in figure 3.1, an electrochemical cell uses a glass vessel containing various components such as a three-electrode assembly consisting of counter, reference, and working electrodes, an electrolyte solution, a glass reservoir, a teflon cap, electrode connections, and an electrolyte solution. The electrolyte solution is a mixture of a solvent and the electrolyte. The liquid solvent dissolves the analyte and supporting electrolyte. It does not lead to any reaction with the electrolyte or analyte and remains stable under oxidation and reduction processes. The supporting electrolyte should also be highly soluble in the solvent. A high electrolyte concentration is important to promote conductivity. Ions from the electrolyte migrate to the electrode to balance the charge and complete the circuit when electron transfer occurs at the electrode. Among the three electrodes, the current flows between the working and counter electrodes, while the reference electrode accurately measures the applied potential. A potentiostat is used to apply the potential to the working electrode with respect to the potential of the reference electrode. A few reference electrodes commonly used in aqueous media include the SHE, the SCE, and the AgCl/Ag electrode. The counter electrode completes the electrical circuit. A current is recorded when electrons flow between the counter and working electrodes. After assembly of the cell and the addition of the analyte, a potential develops at the electrodes. The potential that is present when no current flows is called the open-circuit potential (OCP). The OCP provides information about the redox states of the species in the solution and the concentration of species present in the solution.

The electrolyte solution has an intrinsic resistance in the electrochemical cell. Because of this, the recorded potential may not be due to the analyte in solution, which is called an ohmic drop. The ohmic potential drop in the solution is not an overpotential, because it is not due to the electrode reaction but rather to the bulk solution only. Thus, proper cell design and instrumentation can minimize its contribution to the measured electrode potential. In CV measurements, the effect of an ohmic drop appears as increased separation between peaks in the voltammogram for a redox reaction. To obtain a cyclic voltammogram, the electrodes are connected to a potentiostat, and the experimental parameters are controlled by interfaced software. The basic parameters to be applied are the potential window to be scanned, the scan rate, and the number of scans [9].

3.5 Cyclic voltammetry

The cyclic voltammetry (CV), galvanostatic charge/discharge (GCD), and electrochemical impedance spectroscopy (EIS) are methods used to measure current, voltage, energy over time, power, capacitance/capacity, equivalent series resistance, and the operating potential window. Moreover, parameters such as retention capability, cycling stability, coulombic efficiency (CE), and self-discharge are additional performance metrics of an energy storage system. The testing of three-electrode cells differs from that of two-electrode cells. Among the three main techniques, each provides specific key parameters and hence information. For example, EIS evaluates the ESR, GCD determines the capacitance, energy, and power, while CV differentiates between capacitor-like and battery-like characteristics [2, 7].

In CV measurements, the operating potential window defines the maximum potential range for the chemical stability of the electrolyte under cathodic and anodic reactions and the electrode materials under decomposition by the solvent. The terms potential window and voltage window are used for electrodes and devices, respectively. Moreover, within the potential window, supercapacitors remain reversible during charging and discharging processes. The electrodes must be capacitive. The voltage of a supercapacitor is proportional to the stored charge and reaches its maximum in the fully charged state. Higher capacitances limit the rate of potential change to lower values [10].

Let E_{N1} and E_{N2} denote the lower and upper potential limits of the cathode, respectively, while E_{P1} and E_{P2} denote the lower and upper potential limits of the anode, respectively. Then, the potential window for the cathode is E_{N2}-E_{N1}, while that for the anode is E_{P2}-E_{P1}. Generally, the two windows have an overlap that limits the total operating potential window (figure 3.3). Moreover, if the difference

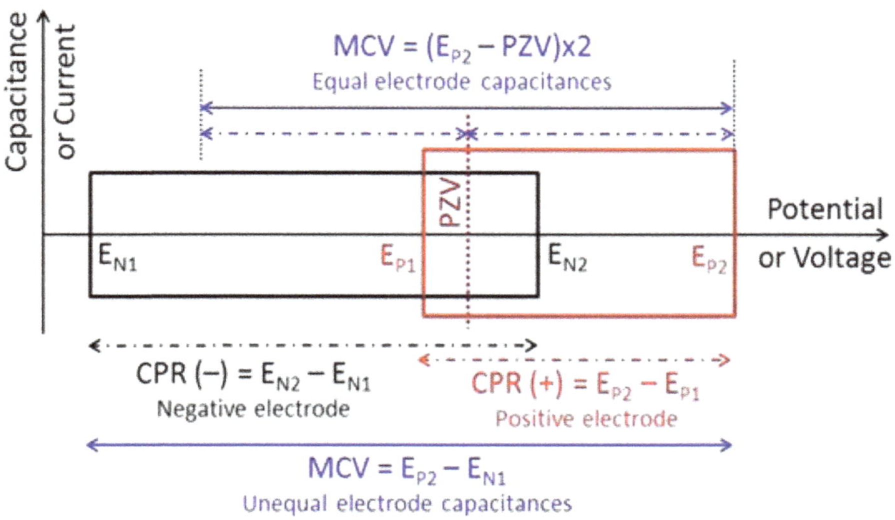

Figure 3.3. Schematic illustration of the operating potential window. Reproduced from [11]. CC BY 4.0.

Table 3.1. Methods of calculating differential and integral capacitances using CV, GCD, and EIS methods.

Capacitance	CV	GCD	EIS
C_{diff}	$\frac{i}{v}$	$\frac{i}{dV/dt}$	$\frac{-1}{\omega Z''}$
C_{int}	$\frac{1}{v(V_{\max}-V_{\min})}\int idV$	$\frac{i\Delta t}{V_{\max}-V_{\min}}$	N.A.

between the potential of zero voltage (PZV) and the upper potential limit of the cathode ($PZV-E_{N2}$) is greater than the difference between the PZV and the lower potential limit of the anode (E_{P1}-PZV), then the anode is the voltage-determining electrode and vice-versa. For electrodes with equal capacitances, the operating voltage window is $2 \times (E_{P2}-PZV)$ if PZV-$E_{N2} > E_{P1}$-PZV or $2 \times (PZV$-$N_{C2})$ if E_{P1}-$PZV > PZV$-E_{N2}. Supercapacitors with unequal capacitances have an operating voltage window of $E_{P2}-E_{N1}$.

Regarding the capacity of batteries and the capacitance of supercapacitors, the capacity is defined by the amount of charge that a battery can deliver at a given voltage rating, while the total capacitance of a supercapacitor is the ratio of the change in charge in response to a change in potential ($C = \Delta q/\Delta V$). The normalized capacitance is defined as the capacitance per unit quantity, x, such as the mass, volume, or area of the electrode active material. For example, $C' = \Delta q/x\Delta V$; however, depending on the normalized quantity, this is called the gravimetric (or specific) capacitance (F g^{1}), the volumetric capacitance (F cm^{3}), or the areal capacitance (F cm^{2}). The differential capacitance of a supercapacitor can be written as $C_{\text{diff}} = dq/dV$ or $C_{\text{diff}} = d(CV)/dV$, and the total static (integral) capacitance is given by $1/V\int_0^V C_{\text{diff}}dV$. CV and GCD measurements provide differential and integral capacitances, but EIS provides only differential capacitance, as shown in table 3.1.

In CV measurements, a linearly varying potential is applied between the cathode and anode in two-electrode systems or between the reference and working electrodes in three-electrode configurations. The sweep rate or scan rate, v (in V s^{-1}), is the speed of the change in potential, and the range of potential change is known as the potential window or operating potential. The electrochemical interaction is monitored by measuring the instantaneous current during the potential sweep at the cathode and the anode. The results show the current (A) with respect to the potential (V). The three-electrode setup helps to distinguish between electric double-layer capacitance (EDLC) and pseudocapacitive capacitance, which have different shapes of cyclic voltammograms. A rectangular CV examines both EDLC and pseudocapacitive behaviors. Some pseudocapacitive materials can also result in reversible redox peaks. The specific capacitance and energy information can be extracted from the CV curves using the integration method mentioned earlier.

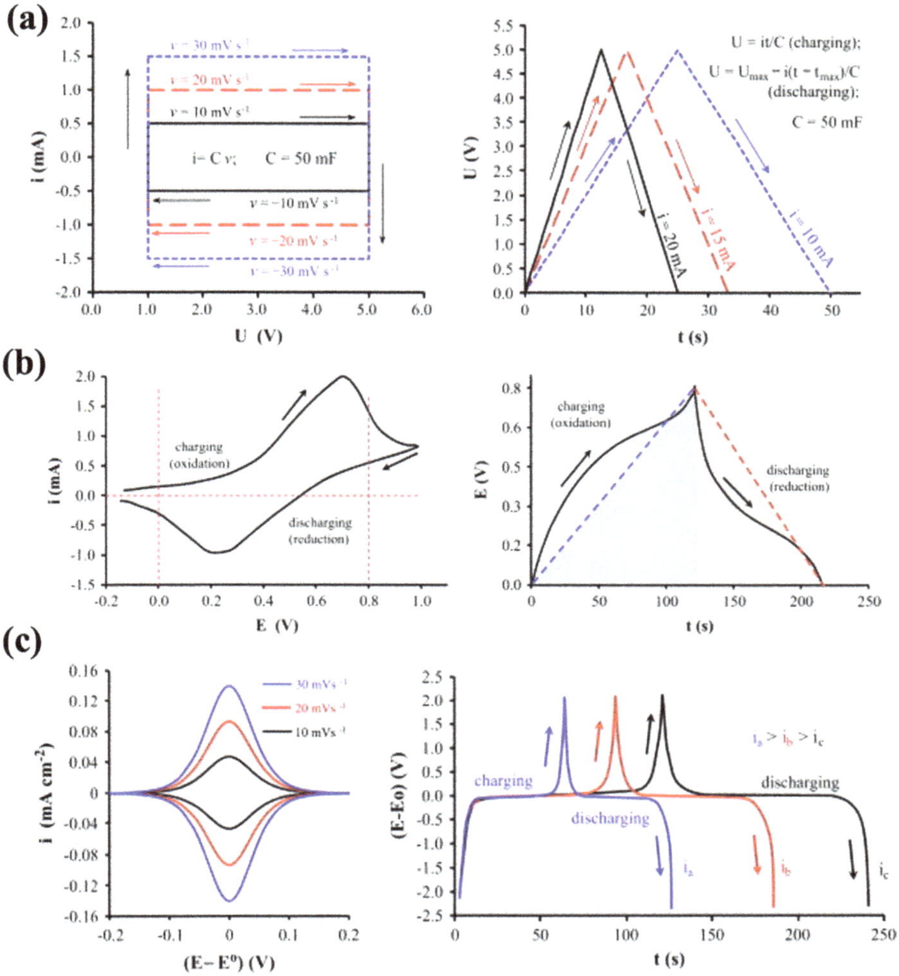

Figure 3.4. CV and GCD profiles of (a) an ideal capacitor. (b) and (c) are of pseudocapacitive supercapacitors. Reprinted from [12], Copyright (2013), with permission from Elsevier.

Figure 3.4 shows the CV and GCD responses of a supercapacitor, highlighting the redox reactions of different species as well as the involvement of adsorbed species on the electrodes [2]. The adsorption of oxidized and redox species can be evaluated from the Nernst equation as follows: $E = E^0 + \frac{RT}{nF}\ln\left(\frac{1-\theta}{\theta}\right)$, where θ is the fractional coverage. The relation is followed by the GCD profile of the adsorbed species or a thin layer of active material coated on the electrode. The area under the CV curve represents the charge required for the full reduction of the adsorbed species. Here, the surface concentration of the adsorbed species is $\Gamma = \frac{q}{nFA}$, and the expression for the current is $i = \frac{n^2F^2A\Gamma\upsilon \exp\left[\frac{nT}{RT}\left(E-E^0\right)\right]}{RT\left\{1+\exp\left[\frac{nT}{RT}\left(E-E^0\right)\right]\right\}^2}$. Figure 3.4(a) shows the

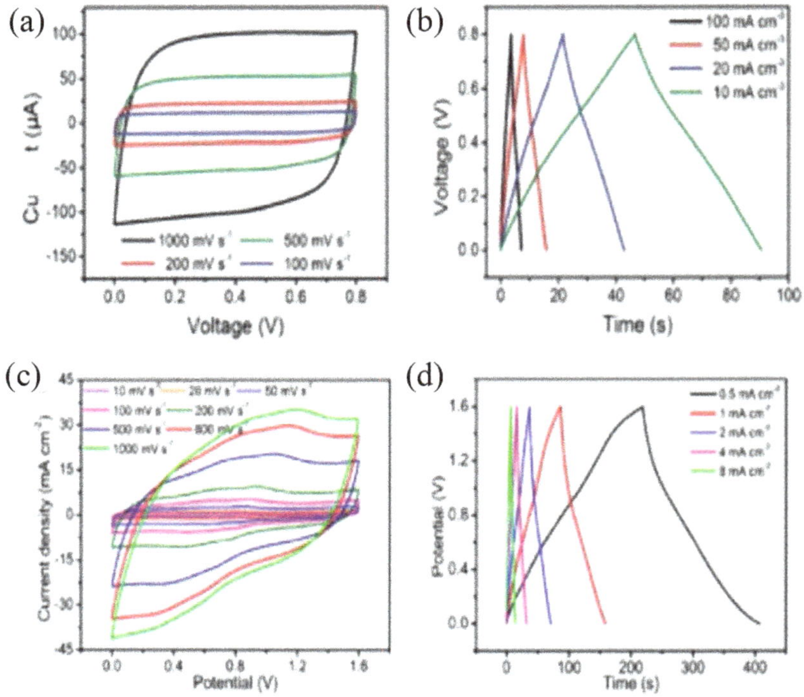

Figure 3.5. Electrochemical responses of supercapacitors made from activated carbon (a) (CV) and (c) (CGD) and modified activated carbon (b) (CV) and (d) (CGD). Reproduced from [13]. CC BY 4.0.

CV results at three different scan rates. In an ideal capacitor, the amount of stored charge on the plates is proportional to the voltage between the plates, which is known as capacitance ($C=q/V$). In CV, a linearly varying potential is applied over time ($V=V_0 + vt$). In this case, the capacitance is given by $C = dq/dV = (dq/dt)/(dV/dt) = i/v$. This relation shows that the current flowing between the plates is independent of the applied potential and linearly proportional to the scan rate, v. The CV (I–V) response of an electric double-layer capacitor results in a rectangular shape due to a positive v when charging and a negative v when discharging. Triangular-shaped GCD results are caused by an increase/decrease in potential at a constant rate, as shown in figure 3.4(a). Although the CVs of EDLCs and pseudocapacitors are different, they both show proportionality between i and v and produce a reverse-polarity current when v is reversed [2], however both the CV and GCD deviate from the ideal capacitor response due to faradic interaction or pseudocapacitance (figure 3.4(b)). Figure 3.4(c) is the response of the battery system. Another example in figure 3.5 depicts the CV and GCD responses of activated carbon and modified activated carbon, showing EDLC and pseudocapacitance.

Upon comparing pseudocapacitance and battery characteristics, it becomes apparent that batteries show less reversible CV and that the oxidation and reduction peak potentials of their voltammograms are shifted toward more positive and negative values, respectively, than those of pseudocapacitors. The GCD profiles of

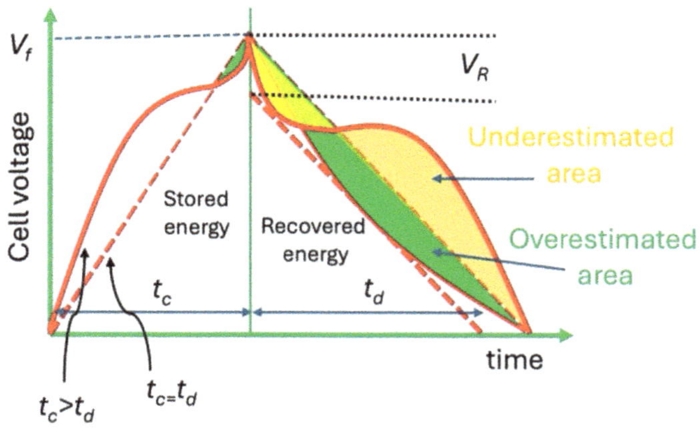

Figure 3.6. Schematic illustration of the GCD shapes of different electrochemical energy storage systems. Discharge profile showing the equivalent series resistance (ESR) of a supercapacitor. GCD profiles of supercapacitor showing considerable deviation from an ideal capacitive behavior.

batteries show plateaus corresponding to oxidation and reduction reactions. As depicted in figure 3.6, this is due to the lower reversibility of their cyclic voltammograms, which results in an upward shift in the charging plateaus and a downward shift upon discharge in a GCD curve.

The CV curves of pseudocapacitive materials are either rectangular in shape or highly reversible with significantly broadened peaks. In batteries, active materials are characterized by distinct peaks in the CV corresponding to redox reactions. In general, the CV response shows the variation in the current in response to an applied potential depending on whether the redox reaction is controlled by diffusion or takes place via an adsorption process at the electrode surfaces. A rectangular shape signifies ideal capacitive behavior. Hybrid systems display the CV characteristics of batteries or supercapacitors in a three-electrode setup. Hybrid and pseudocapacitor systems typically have cyclic voltammograms that are neither sharply peaked nor perfectly rectangular. Instead, they are quasi-rectangular with humps corresponding to the redox peaks; thus depicting the combined behaviors of batteries and supercapacitors. It is not possible to identify battery-like or supercapacitive behavior from such CV curves. Ardizzone *et al* provided a way to quantify the contributions of diffusion-controlled and capacitive charge storage processes [14]. The dependency of the stored charge $q(v)$ can be evaluated by integrating the internal area of the CV over the scan rate as follows: $q(v) = q_{\text{capacitive}} + k_3 v^{-1/2}$. The charge, $q_{\text{capacitive}}$, is due to the contribution from the outer surface of the electrode material, which is the area directly accessible to the electrolyte. The second term of the total charge is the inner voltammetric charge, contributed by partially accessible sites such as pores, grain boundaries, cracks, etc. which is proportional to $v^{-1/2}$. The scan rate in CV measurements controls the diffusion rate of the ionic species in the bulk. As $v \to \infty$ (at very high scan rates), the faradic charge storage processes are too low, while for $v \to 0$, diffusion-controlled processes become dominant by reaching all the pores inside the bulk of the electroactive electrode material. In another analysis, the process of two mechanisms was written as $i = i_{sc} + i_{bat.} = k_1 v + k_2 v^{-1/2}$ or $i/v^{-1/2} = k_1 v^{-1/2} + k_2$, where for a supercapacitive process, the current response varies directly with v, while for a battery-type material, it is $v^{-1/2}$. The values of k_1 and k_2 can be obtained from the slope and intercept of a plot of $i/v^{-1/2}$ vs. $v^{-1/2}$ [7].

The capacitance of a cell can be obtained from the CV curve using the following relation:

$$C_s = \frac{\int_{V_i}^{V_f} idV}{v \times \Delta V \times x},$$

where i is the current (in A) and $\int idV$ is the volumetric charge obtained by integrating the area under the reverse peak of the CV with dimensions of amp-volts. v is the scan rate (V s^{-1}), ΔV is the potential interval (from V_i to V_f and again from V_f to V_i), and x can denote the mass, volume, or surface area of either electrode or the whole device. In an ideal case, the area under the forward or backward scan can be selected, or the integrated area over the entire potential range can be used, divided by two. However, practically, the area under the forward scan is greater than that of the reverse scan. Hence, the capacitance should be calculated based on the area under the reverse scan of the CV. Figure 3.7 shows different manifestations of non-ideality in the CV response [15].

The equivalent series resistance (ESR) is composed of the electrolyte resistance, the resistance of the electrode material, and the contact resistance between the electrode material and the current collector. The presence of the ESR lowers the current and rounds off the two corners of the cyclic voltammogram at the beginnings of the charge and discharge processes. Therefore, sharper curves represent lower resistances. Figure 3.7(a) shows the CV response of an ideal capacitor; however, figure 3.7(b) depicts a deviation from this response that corresponds to the CV profile of sloping curve, where the slope is proportional to the faradic resistance of the system ($1/R$). The circuit shows the faradic charge transfer process in parallel to double-layer charging. A simple pseudocapacitive circuit is also presented in figure 3.7(c). The resistance of the RC circuit (equivalent to the ESR) can be obtained by $i = Cv\ (1 - e^{-t/RC})$, where C is the capacitance, v is the potential sweep rate (in V s^{-1}), and R is the resistance of the RC circuit. The equation can be fitted to the initial part of the CV to obtain the value of the ESR [16].

Figure 3.8 shows the effect of deviation from the rectangular CV and triangular GCD plots at high scan rates. It illustrates the increase in internal resistance caused by the involvement of both the double-layer and faradic (pseudocapacitive) charge storage mechanisms between the electrolyte and the active materials.

The energy and power of an electrode or a device can be calculated from the capacitance. Alternatively, the maximum specific power (W kg^{-1}) or power density (P) (W x^{-1}) can be calculated from the initial discharge process using the following relationship: $P = \frac{V^2}{4ESRx}$, where V is the potential (in V), ESR is the series resistance (in ohms), and x is the volume, mass, or surface area of the electrode material or the package. Figure 3.9 shows a plot of both the specific power and cell voltage versus current density. The specific power initially increases with the increase in the current

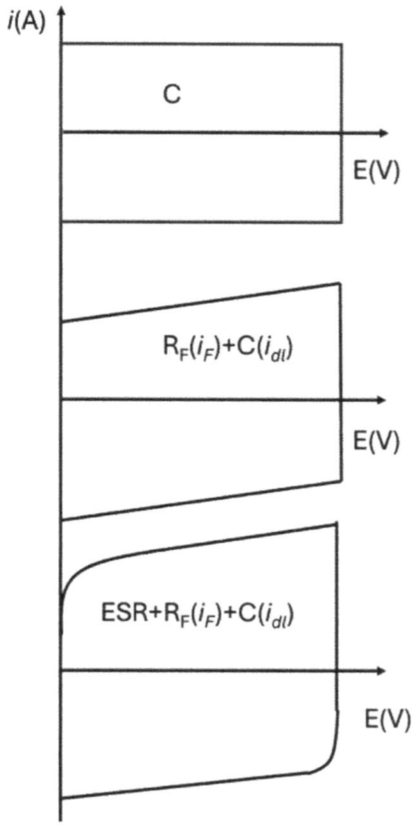

Figure 3.7. Schematic representation of the equivalent circuit model, showing the corresponding CV curves of (a) an ideal electrochemical capacitor, (b) a capacitor with a faradic charge-transfer process, and (c) a simplified supercapacitor model that includes an ESR, a faradic resistance, and a double-layer capacitor.

density and then starts decreasing at higher current densities after reaching a maximum value at nearly half of the discharging cell voltage. Another parameter, rate capability, measures capacitance as a function of the scan rate. The highest specific capacitance is obtained at the lowest scan rate, and it reduces at higher scan rates. The penetration depth of the electrolyte ions inside the electrode materials drastically reduces with increasing scan rates because the inner portions of the materials no longer play a significant role in the charge storage process [18].

The ESR includes the resistance of the bulk electrolyte, the contact resistance between the electrode material and the current collector, and the resistance of the electrode. The ESR limits the charging and discharging performance of the supercapacitor, thus, lower ESR values lead to high performance supercapacitors.

3.6 Galvanostatic charge–discharge analysis

GCD allows one to analyze the detailed energy storage performance of electrode materials. A constant current is applied, and the electrode is charged and discharged

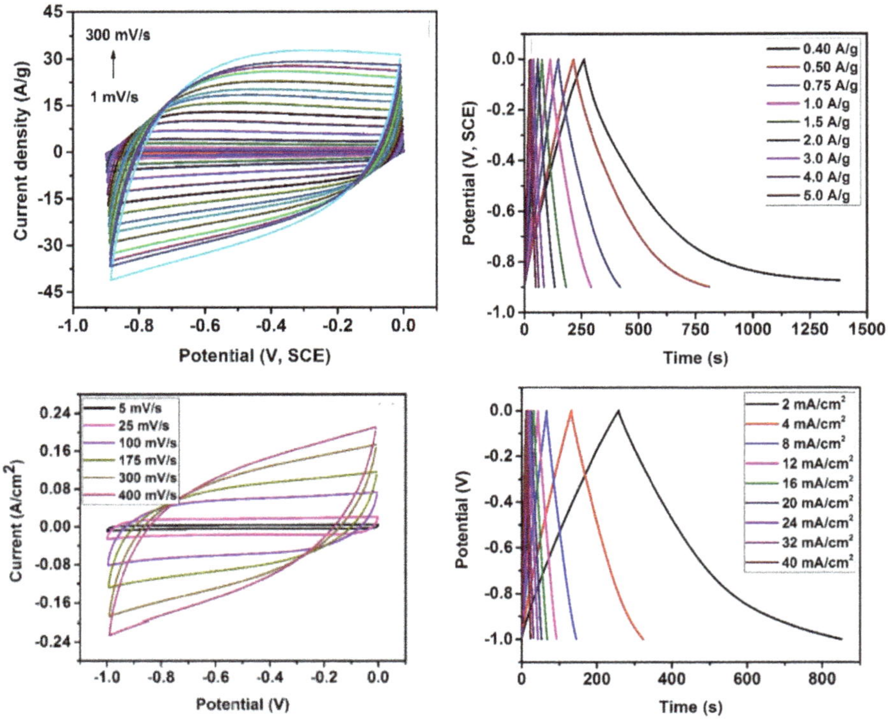

Figure 3.8. The effect of ESR (IR drop) on the shape of CV curves and measured specific capacitance as a function of scan rate. Reproduced from [16]. CC BY 4.0.

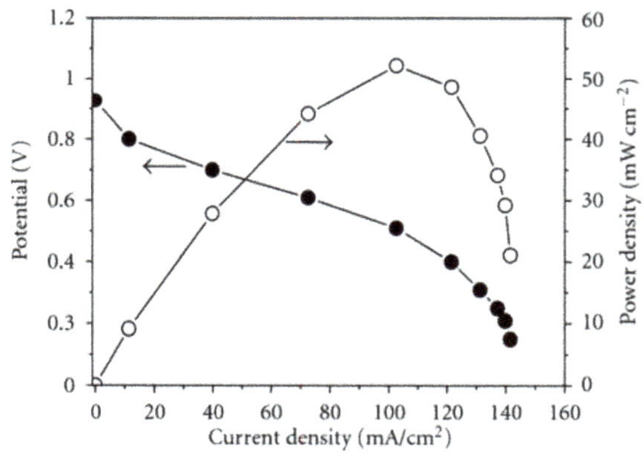

Figure 3.9. Power density and cell potential as a function of current density. The measurements were made using proton exchange membrane fuel cells. Reproduced from [18]. CC BY 3.0

between two potential limits. The changing voltage over time is recorded for analysis. The GCD output (voltage) of EDLC and some pseudocapacitive materials increases and decreases linearly with the applied current. However, hybrid systems and some pseudocapacitive materials show nonlinear charge–discharge responses,

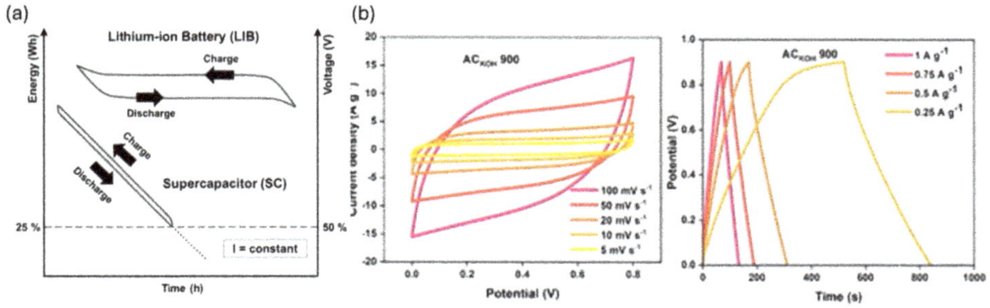

Figure 3.10. (a) GCD profiles of a battery, an EDLC, and a hybrid device. The profiles of pseudocapacitors resemble those of hybrid energy storage devices. The area under the discharge profile of a device represents its energy level. Reproduced from [19]. CC BY 4.0. (b) CV and GCD profiles of activated carbon at different potential windows and current densities, respectively. Reprinted with permission from [20]. Copyright (2024) American Chemical Society.

possibly accompanied by voltage plateaus. Batteries show a long, steady voltage plateau due to thermodynamic conditions (figure 3.10(a)). The GCD response is affected by the current density in much the same way that the scan rate affects the CV response. The surface only contributes to the charge–discharge processes at high current densities. Lower current densities lead to high energy storage due to the accessibility of inner and outer sites to the electrolyte ions. The cell voltage should slowly be increased until it reaches its peak value. In a high cell voltage region, the appearance of a plateau response shows that current is consumed in decomposing the electrolyte. As discussed later, the coulombic efficiency (CE) decreases when the voltage window is increased beyond the stability limit of the electrolyte, as shown in figure 3.10(b) [2].

The capacitance can also be calculated using the GCD discharge curve. The capacitance of an EDLC process can be evaluated from the linear variation of voltage over time during the constant current discharge process using $C = \frac{\Delta q}{\Delta V} = \frac{i\Delta t}{\Delta V}$. In a hybrid or pseudocapacitive system, where the GCD response is not a perfect triangle, the capacitance is given by $C = i/(dV/dt)$. A general equation used to evaluate the q/V value from the discharge curve (linear or nonlinear), which is not a capacitance term in a hybrid system, is $\frac{q}{V} = 2i\int Vdt/xV^2 |_{V_i}^{V_f}$, where q is the amount of charge, V is the voltage range in V, i is the discharge current in A, and the normalizing factor x can represent the mass in g, the volume in cm^3, or the area in cm^2 of the electrode material. For a normalized current in the above expression, x should not be considered in the denominator of the equation. For pure EDLC, i.e. a linear GCD profile, $q/V = it/V = C$, where t is the discharge duration in s. If there is an IR drop, then $C = I\Delta t/(\Delta V - \mathrm{ESR})$.

In addition to ESR, some supercapacitive systems exhibit an internal resistance caused by the redistribution of charges in the pores (equivalent distributed resistance, EDR). The EDR appears as a curvature after the ESR has been accounted for. Only the linear portion of the equation should be used to calculate the capacitance, as follows: $C = I(t_2 - t_1)/(V_2 - V_1)$. As mentioned earlier for nonlinear GCD, the capacitance concept should not be applied; instead, energy storage capacity or energy should be used. The energy/capacitance between the positive and negative electrodes should not be measured from the negative electrode in a three-electrode system, as shown in figure 3.1. Mass loading is a critical factor for commercial applications, and both mass loading and thickness are optimized in devices. Thinner electrodes may be better for low-power applications, while thicker electrodes are better for high-power applications. The capacitance obtained from thinner materials is higher, but such materials are often practically not usable. The gravimetric energy/capacitance decreases as the mass loading increases, and the areal performance increases linearly with the mass loading. Thus, to avoid overestimating supercapacitor performance, comparisons of supercapacitors should consider the mass loading, volume, area, and thickness of the active materials required to obtain comparable performance. During charge–discharge processes, the energy of a supercapacitor can be evaluated from the relation $E = \int_0^q Vdq = i\int_0^t dt$; however, for linear GCD, this can be simplified to $E = 1/2CV^2$. Energy and power are the key performance parameters of supercapacitor devices. An incorrect interpretation of capacitance should be avoided by using $C = \frac{\Delta q}{\Delta V}$ from a nonlinear GCD profile rather than from a peak-shaped cyclic voltammogram to calculate $E = 1/2CV^2$. In such cases, the integral of the area under the GCD graph should be used. Although CV and GCD are used for both two-electrode and three-electrode systems, the power and energy can only be measured using two-electrode systems and not three-electrode setups. The value of the ESR or IR drop can be calculated from the initial part of the discharge profile as follows: $\text{ESR} = \Delta V_{\text{ESR}}/|I_{\text{charge}}| + |I_{\text{discharge}}| = \frac{\Delta V_{\text{ESR}}}{2I}$, while the value of the EDR can be obtained from the discharge curve using $\text{EDR} = \Delta V_{\text{EDR}}/|I_{\text{charge}}| + |I_{\text{discharge}}| = \frac{\Delta V_{\text{EDR}}}{2I}$.

The rate capability of an energy storage device can be evaluated at different current densities. Devices with stable capacitance at each current density (such as those made using EDLC materials) are known for their high rate capability [22]. However, pseudocapacitive materials suffer from high internal resistance due to faradic redox reactions; thus, they exhibit lower rate capability.

Cycling stability is another critical performance parameter which can be measured for supercapacitors by continuously repeating GCD cycles at a given current density for several thousand cycles. However, for batteries, such testing is limited to just a few hundred cycles. The percentage of its initial capacitance that a device retains after a specified number of cycles is usually evaluated from cyclic tests. The CE quantifies the efficiency of electron transfer within an electrode or device. It is calculated as follows:

$$\text{CE}(\%) = \Delta q_d/\Delta q_c \times 100 = \frac{\int_0^{t_d} i_d dt}{\int_0^{t_c} i_c dt \times 100},$$

where q_d and q_c are the charges extracted and inserted into the electrode materials, respectively; t_d and t_c are discharging and charging times, respectively; and i_d and i_c

are the discharging and charging currents, respectively. The relation can be simplified if charging and discharging occur at the same current, as follows: $\mathrm{CE}(\%) = t_d/t_c \times 100$. The CE is influenced by current density and temperature; thus, both quantities should be mentioned in reports. The energy efficiency (EE) can be obtained from $\mathrm{EE}(\%) = E_d/E_c \times 100 = \frac{\int_0^{t_d} i_d V_d dt}{\int_0^{t_c} i_c V_c dt \times 100}$, where E_d and E_c are the discharging and charging energies, i.e. the areas under the discharge and charge profiles, respectively. The voltage efficiency (VE) can be obtained using $\mathrm{VE}(\%) = (\mathrm{EE}/\mathrm{CE}) \times 100$ [23].

3.7 AC impedance analysis

The abovementioned measurement techniques impose potential sweeps, potential steps, or current steps, where the response of the working electrode is measured after perturbing the electrode condition far from equilibrium. These are usually transient signals. Steady-state measurements can be taken by perturbing the system with a small alternating signal. Steady-state measurements provide high precision, linearized current–potential characteristics, and the ability to measure responses over a wide time (frequency) range. In faradic impedance measurements, the cell is inserted as an unknown impedance to form one arm of an impedance bridge, and the bridge is balanced by adjusting R and C in the opposite arm of the bridge. The values of equivalent resistance and capacitance that are connected in series can be determined as a function of the variation in the cell in response to frequency, where the electrode impedance is plotted versus frequency, which is called EIS. In AC voltammetry, a three-electrode cell is used, and a DC mean value of the potential, E_{dc}, and a small (~5 mV) sinusoidal potential, E_{ac} (peak-to-peak amplitude), are applied at the working electrode. The output responses are the magnitude of the AC component of the current at the given frequency and the phase angle with respect to E_{ac}. The role of the DC component is to set the mean surface concentration at the electrode, and since E_{dc} remains steady, the thickness of the layer becomes thick enough to exceed the diffusion zone in the electrolyte, which is sensitive to the perturbation introduced by E_{ac}. Hence, an electrochemical cell represents an impedance to a small sinusoidal excitation in an equivalent circuit of resistors and capacitors that passes current with a small amplitude and phase angle. These circuits are called Randles equivalent circuits, as shown in figure 3.11 [24].

The parallel elements depict that the total current passing through the working electrode is the sum of the contributions from the faradic process, i_f, and double-layer charging, i_c. The double-layer capacitance is the same as the electrostatic capacitance that we observed earlier; thus, in the equivalent circuit, it is represented by C_d. The faradic process is represented by an impedance, Z_f, unlike the linear elements (R and C) that work independently of frequency. A series element, R_Ω, is considered for the solution resistance. Two equivalent circuits can be considered for the faradic impedance. The faradic impedance is represented by the series combination of the resistance of the electrolyte, R_s, and the pseudocapacity, C_S; alternatively, it can be represented by the series combination of a pure resistance, R_{ct}, and the Warburg

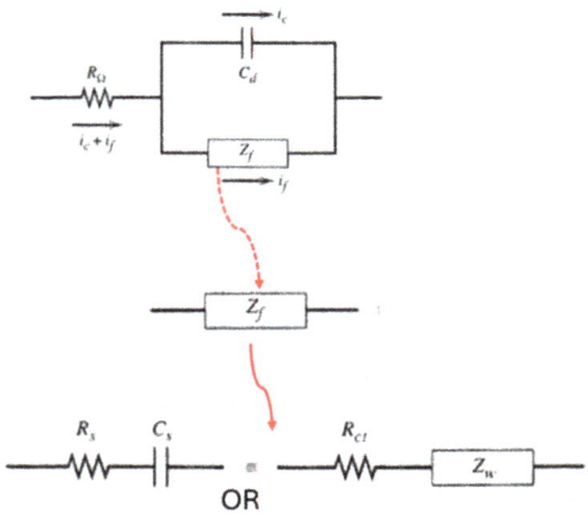

Figure 3.11. Schematic diagram of (a) an equivalent circuit representing an electrochemical cell and (b) the subdivision of Z_f into R_s and C_s, or into R_{ct} and Z_w.

impedance, Z_w, which represents a resistance to mass transfer. Hence, the elements R_Ω and C_d are ideal circuit components, while the faradic impedance changes with frequency, ω. Thus, faradic impedance measurements mainly evaluate the frequency dependencies of R_s and C_s. The equivalent circuits for electrochemical cells are not unique and vary with the complicated processes.

Usually, an electrode process at the electrode–electrolyte interface is a heterogeneous reaction. The rate of reaction depends on the mass transfer to the electrode and other surface effects while the usual kinetics takes place and is given by rate (mol^{-1} cm^{-2}) $= \frac{i}{nFA} = \frac{j}{nF}$, where j is the current density (A cm^{-2}). The features of the i–E curve provide insight into the electrode reaction. The term polarization is defined as the shift in electrode potential from the equilibrium potential when a faradic current is applied. The overpotential, η, is defined as $\eta = E - E_{\mathrm{eq}}$. The ideal polarizability is a horizontal region of an i–E curve, and an ideal nonpolarizable electrode is one where the potential of the electrode remains constant. A cell composed of two ideal nonpolarizable electrodes has an i–E response similar to a pure resistance [25]. The current in the electrode is governed by the rate of processes including mass transfer (from the electrolyte to the electrode surface), electron transfer at the electrode surface, chemical reactions on the electrode surface, and other surface processes, such as adsorption, desorption, or electrodeposition, as shown in figure 3.12.

The current in each process is driven by a certain overpotential η, which is the sum of terms relevant to the different steps of the total electrode process, such as the mass transfer overpotential, η_{mt}, the charge transfer overpotential, η_{ct}, the overpotential associated with the electrode reactions, η_{rxn}, etc. The electrode reaction can be represented by a resistance R composed of a series of resistances (or impedances) in various steps such as R_{mt}, R_{ct}, etc. as shown in figure 3.13.

The mass transfer and charge transfer resistances become prominent at high current densities. The charge transfer resistance, R_{ct} during the passage of current, a

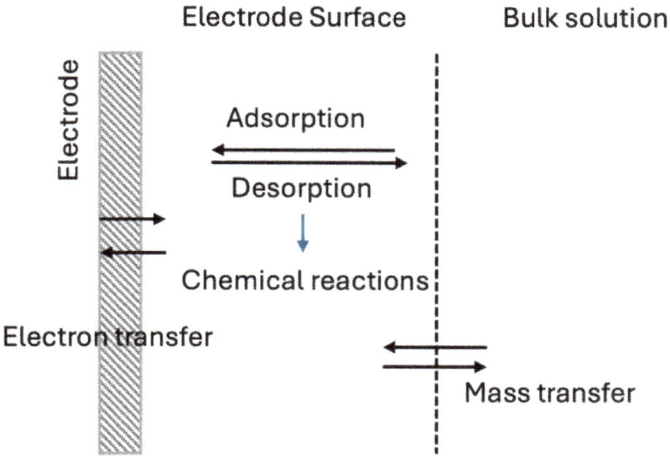

Figure 3.12. A general electrode reaction and processes.

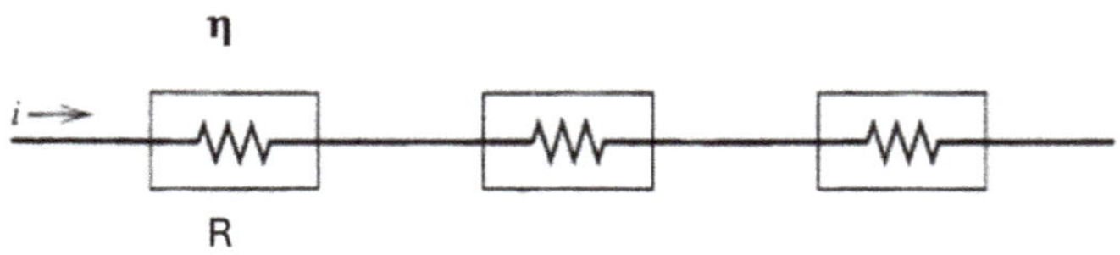

Figure 3.13. Processes in an electrode reaction represented as resistances.

voltage drop, iR_s, occurs, where R_s is the solution resistance (this also behaves as a true resistance) between the two electrodes for a wide range of conditions. The ohmic potential drop in the electrolyte is not a form of overpotential, as it is a characteristic of the electrolyte rather than the electrode reaction. The applied potential, E_{appl}, of the working electrode, is thus given by E_{appl} (vs. SCE) $= E_{eq}$ (vs. SCE) $+ \eta - iR_s$, where E_{eq} is the equilibrium value of the potential at the electrode.

The total voltage drop as an AC current passes through the impedance is $E = iR_s + \frac{q}{C_S}$; therefore, $\frac{dE}{dt} = R_S\frac{di}{dt} + \frac{i}{C_S}$. If the current is $i = I \sin \omega t$, then $\frac{dE}{dt} = (R_s I\omega t)\cos \omega t + \left(\frac{I}{C_S}\right)\sin \omega t$, which provides a link between and R_s and C_s that is given by $R_s = R_{ct} + \sigma/\omega^{1/2}$ and $C_s = \frac{1}{\sigma\omega^{1/2}}$. The terms $\sigma/\omega^{1/2}$ and $\frac{1}{\sigma\omega^{1/2}}$ contribute to mass transfer. The faradic impedance can be divided into the charge transfer resistance, R_{ct}, and the Warburg impedance, Z_w, as shown in figure 3.11. R_{ct} is defined as a negative ratio, $-\frac{\eta}{i}$, given that the overpotential occurs in a narrow potential range. The Warburg impedance is a frequency-dependent resistance given by $R_w = \sigma/\omega^{1/2}$, which is in series with the pseudocapacitance, $C_w = C_s = \frac{1}{\sigma\,\omega^{1/2}}$. Thus, the total faradic impedance, Z_f, is given by $Z_f = R_{ct} + R_w - j/(\omega C_w) = R_{ct} + [\sigma\omega^{-1/2} - j(\sigma\omega^{-\frac{1}{2}})]$. The terms within the bracket are known as the Warburg impedance. For a standard chemical system of a one-electron process, $O + ne \rightleftharpoons R$,

$E = E[i,\ C_o(0,\ t),\ C_R(0,\ t)]$, the concentrations at the electrode surfaces and in the bulk solution are denoted by $C_o(x = 0)$ and C_o*, respectively. For linear diffusion, $C_o(x,0) = C_o*$ and $C_R(x,0) = C_R*$. For small overpotentials, the relation between i and η can be written as

$$\frac{i}{i_0} = \frac{C_o(0,\ t)}{C_o^*} - \frac{C_R(0,\ t)}{C_R^*} - \frac{F\eta}{RT} \text{ and } \eta = -i(R_{ct+}R_{mt,\ c} + R_{mt,\ a}),$$

where the terms for charge and mass pseudoresistances are defined as, for example, $R_{mt} = \frac{RT}{nF\mid i_l \mid}$, $R_{ct} = \frac{RT}{nFi_0}$, and $R_s - \frac{1}{\omega C_S} = R_{ct} = \frac{RT}{Fi_0}$, where i_o and i_l are the exchange and limiting currents, respectively. The limiting current is defined as the current flowing when the electrode process occurs at the maximum rate of mass transfer when $C_o(x = 0) = 0$. Thus, the limiting current is $i_l = nFAm_oC_o^*$. The exchange current is defined by the magnitude of either current, i_c or i_a, when the net current is zero at equilibrium. The exchange current, $i_o = FAk^0C_o^{*(1-\alpha)}C_R^{*\alpha}$, where k^0 is a standard rate constant. For $C_o^* = C_R^* = C$, $i_o = FAk^0C$, where α is a transfer coefficient, which is a measure of the symmetry of the energy barrier. In a reversible system, when the charge transfer kinetics are very fast (as in the limiting case, $i_o \to \infty$), then $R_{ct} \to 0$ and $R_s \to \sigma/\omega^{1/2}$, which express the relationship between resistance and frequency. The slope is given by

$$\sigma = \frac{RT}{n^2F^2A\sqrt{2}}\left(\frac{1}{D_o^{1/2}C_o^*} + \frac{1}{D_R^{1/2}C_R^*}\right).$$

In general, D_i is the diffusion coefficient (in cm^2 s) and C_i the concentration of species i (in mol cm^{-3}).

An intercept of the line R_{ct} in the plot of R_s gives the exchange current i_o. The Warburg impedance becomes negligible at high frequencies because the diffusion-induced current becomes less significant. The reason for this is that within short timescales, diffusion reduces significantly [26].

The total impedance of the equivalent circuit given in figure 3.11 is $Z(\omega) = R - j/(\omega C) = Z_{Re} - jZ_{Im}$. This varies with frequency, where Z_{Re} and Z_{Im} are the real and imaginary parts of the impedance, e.g. $Z_{Re} = R$ and $Z_{im} = XC = 1/\omega C$. The magnitude of $\boldsymbol{Z}$ is given by $Z^2 = R^2 + X_C{}^2 = (Z_{Re})2 + (Z_{Im})^2$, and the phase angle, φ, is given by $\tan\varphi = Z_{Im}/Z_{Re} = X_C/R = 1/\omega C$. For a pure resistance, $\varphi = 0$, and for a pure capacitance, $\varphi = \pi/2$. The variation of impedance with frequency can be displayed in a Bode plot, where both log $|Z|$ and φ are plotted against log ω; alternately, a Nyquist plot uses Z_{Im} vs. Z_{Re} for different values of ω, as shown in figure 3.14.

Involving electrochemical species in calculations does not help in direct measurements of nonfaradic impedance. Thus, to obtain the total impedance as a function of ω, we directly evaluate the faradic impedance R_Ω and the double-layer capacitance C_d. Referring to the equivalent circuit in figure 3.11, the real part of the impedance is given by

$$Z_{\mathrm{Re}} = R = R_\Omega + \frac{R_s}{A^2 + B^2},$$

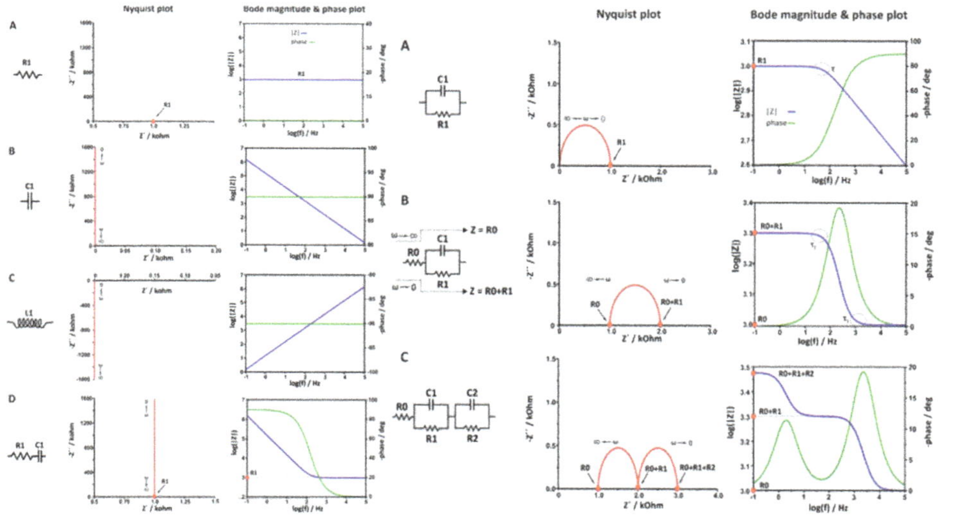

Figure 3.14. Representative Bode and Nyquist plots of example circuits. Reprinted with permission from [26]. Copyright (2023) American Chemical Society.

where $A = (C_d/C_s) + 1$, $B = \omega R_s C_d$, and $Z_{Im} = \frac{1}{\omega C_B} = \frac{B^2/\omega C_d + \frac{A}{\omega C_S}}{A^2 + B^2}$. Following the substitution of the R_s and C_d values, $Z_{\text{Re}} = R_\Omega + \frac{R_{ct} + \sigma\omega^{-1/2}}{(C_d\sigma\omega^{\frac{1}{2}} + 1)^2 + \omega^2 C_d^2 (R_{ct} + \sigma\omega^{-\frac{1}{2}})^2}$ and $Z_{Im} = \frac{\omega C_d (R_{ct} + \sigma\omega^{-\frac{1}{2}})^2 + \sigma\omega^{-\frac{1}{2}}(C_d\sigma\omega^{\frac{1}{2}} + 1)}{(C_d\sigma\omega^{\frac{1}{2}} + 1)^2 + \omega^2 C_d^2 (R_{ct} + \sigma\omega^{-\frac{1}{2}})^2}$. Plots of Z_{Im} and Z_{Re} for different values of ω provide chemical information. We now consider the limiting case of low frequency. When $\omega \longrightarrow 0$, $Z_{\text{Re}} = R_\Omega + R_{ct} + \sigma\omega^{-1/2}$ and $Z_{Im} = \sigma\omega^{-1/2} + 2\sigma^2 C_d$. The elimination of ω gives $Z_{Im} = Z_{\text{Re}} - R_\Omega - R_{ct} + 2\sigma^2 C_d$; thus, plots of Z_{Re} and Z_{Im} should be linear with a unity slope. The extended intercept at the real axis gives $R_\Omega + R_{ct} - 2\sigma^2 C_d$. Hence, the linear correlation between Z_{Re} and Z_{Im} is due to a diffusion-controlled electrode process caused by the frequency-dependent Warburg impedance. The charge transfer resistance R_{ct} and double-layer capacitance C_d become dominant as the frequency increases, and the Warburg impedance becomes unimportant.

The impedance is $Z = R_\Omega - j\left(\frac{R_{ct}}{R_{ct} C_d \omega - j}\right)$, where $Z_{\text{Re}} = R_\Omega + \frac{R_{ct}}{1 + \omega^2 C_d^2 R_{ct}^2}$ and $Z_{Im} = \frac{\omega C_d R_{ct}^2}{1 + \omega^2 C_d^2 R_{ct}^2}$. The elimination of ω leads to the following equation $\left(Z_{\text{Re}} - R_\Omega - \frac{R_{ct}}{2}\right)^2 + Z_{Im}^2 = \left(\frac{R_{ct}}{2}\right)^2$, which is the equation for a circle between Z_{Re} and Z_{Im} at $Z_{\text{Re}} = R_\Omega - \frac{R_{ct}}{2}$ and $Z_{Im} = 0$, with a radius of $\frac{R_{ct}}{2}$. The imaginary component of the impedance is contributed by C_d only, and at high frequencies, it does not offer any impedance, so its contribution becomes zero. At this point, the current and impedance are the charging current and the ohmic resistance. As the frequency decreases, Z_{Im} becomes the finite impedance of C_d, and at very low frequencies, C_d offers high impedance. Current now passes through R_{ct} and R_Ω only. At this point, Z_{Im} falls to zero again, and the Warburg impedance becomes dominant at this frequency [28]. Combining the two features of the limiting cases provides an impedance plot in the complex plane.

References

[1] Guo Q, Sun C, Li Y, Li K and Tai X 2025 Recent Advances of Electrode Materials Applied in an Electrochromic Supercapacitor Device *Molecules* **30** 182

[2] Sharma S and Chand P 2023 Supercapacitor and electrochemical techniques: A brief review *Results Chem.* **5** 100885

[3] Schoetz T, Gordon L W, Ivanov S, Bund A, Mandler D and Messinger R J 2022 Disentangling faradaic, pseudocapacitive, and capacitive charge storage: A tutorial for the characterization of batteries, supercapacitors, and hybrid systems *Electrochim. Acta* **412** 140072

[4] Sun Y 2024 On a New Point of View Toward the Mechanism of Those Electrochemical Reactions Driven by the External Potential *Sci. Front.* **5** 123–9

[5] Ramón J E, Martínez-Ibernón A, Gandía-Romero J M, Fraile R, Bataller R, Alcañiz M, García-Breijo E and Soto J 2019 Characterization of electrochemical systems using potential step voltammetry. Part I: Modeling by means of equivalent circuits *Electrochim. Acta* **323** 134702

[6] Bieniasz L K, González J, Molina Á and Laborda E 2010 Theory of linear sweep/cyclic voltammetry for the electrochemical reaction mechanism involving a redox catalyst couple attached to a spherical electrode *Electrochim. Acta.* **56** 543

[7] Pholauyphon W, Charoen-amornkitt P, Suzuki T and Tsushima S 2024 Perspectives on accurately analyzing cyclic voltammograms for surface- and diffusion-controlled contributions *Electrochem. Commun.* **159** 107654

[8] Montella C 2017 Further investigation of the equivalence of staircase and linear scan voltammograms. II - Effects of electron transfer kinetics, Ohmic drop and double-layer charging *J. Electroanal. Chem.* **799** 194

[9] Elgrishi N, Rountree K J, McCarthy B D, Rountree E S, Eisenhart T T and Dempsey J L 2018 A practical beginner's guide to cyclic voltammetry *J. Chem. Educ.* **95** 197–206

[10] Lim J M *et al* 2023 Advances in high-voltage supercapacitors for energy storage systems: materials and electrolyte tailoring to implementation *Nanoscale Adv.* **5** 615

[11] Dai Z, Peng C, Chae J H, Ng K C and Chen G Z 2015 Cell voltage versus electrode potential range in aqueous supercapacitors *Sci. Rep.* **5** 9854

[12] Chen G Z 2013 Understanding supercapacitors based on nano-hybrid materials with interfacial conjugation *Prog. Nat. Sci.: Mater. Int.* **23** 245–55

[13] Joseph K, Kasparian H and Shanov V 2022 Carbon Nanotube Fiber-Based Wearable Supercapacitors—A Review on Recent Advances *Energies* **15** 6506

[14] Ardizzone S, Fregonara G and Trasatti S 1990 "Inner" and "outer" active surface of RuO_2 electrodes *Electrochim. Acta* **35** 263–7

[15] Kulova T L and Skundin A M 2020 Cyclic voltammetry of supercapacitors with the simplest equivalent circuit *Russ. Chem. Bull.* **69** 1672

[16] Zhang Y, Li X, Li Z and Yang F 2024 Evaluation of electrochemical performance of supercapacitors from equivalent circuits through cyclic voltammetry and galvanostatic charge/discharge *J. Energy Storage* **86** 111122

[17] Zequine C, Ranaweera C K, Wang Z, Singh S, Tripathi P and Srivastava O N 2016 High Per formance and Flexible Supercapacitors based on Carbonized Bamboo Fibers for Wide Temperature Applications *Sci. Rep.* **6** 31704
[18] Yunita A, Farma R, Awitdrus A and Apriyani I 2023 The effect of various electrolyte solutions on the electrochemical properties of the carbon electrodes of supercapacitor cells based on biomass waste *Mater. Today Proc.* **87** 246
[19] Lin R, Qiao J, Zhang H, Cao C, Fu J and Ma J 2011 Electrooxidation of Hydrogen on Ni-Organic Metal Complex Catalysts in Acidic Media for PEMFCs *J. Nanomater.* **2011** 1–9
[20] Kim K, An J, Park K, Roh G and Chun K 2019 Analysis of a Supercapacitor/Battery Hybrid Power System for a Bulk Carrier *Appl. Sci.* **9** 1547
[21] Tomy M, Geethamma G S, Aravind A M, Reshmi S S and Suryabai X T 2024 Effect of Activation Environment on Coconut-Husk-Derived Porous Activated Carbon for Renewable Energy Storage Applications *ACS Sustain. Resour. Manag.* **1** 1534–47
[22] Hardianto Y P, Shah S S, Shuaibu A D, Mohamed M, Sarker S, Alzahrani A S and Aziz M A 2024 *Electrochim. Acta* **513** 145552
[23] Liu Y, Xu N, Chen W, Wang X, Sun C and Su Z 2018 Supercapacitor with high cycling stability through electrochemical deposition of metal–organic frameworks/polypyrrole positive electrode *Dalt. Trans.* **47** 13472
[24] Scholz F 2015 Voltammetric techniques of analysis: the essentials *ChemTexts* **1** 17
[25] Zhang J, Wang S and Ono K 2023 *Microscopy and Microanalysis for Lithium-Ion Batteries* (Boca Raton, FL: CRC Press) 11
[26] Cruz-Manzo S and Greenwood P 2020 An impedance model based on a transmission line circuit and a frequency dispersion Warburg component for the study of EIS in Li-ion batteries *J. Electroanal. Chem.* **871** 114305
[27] Lazanas A C and Prodromidis M I 2023 Electrochemical Impedance Spectroscopy—A Tutorial *ACS Meas. Sci. Au* **3** 162–93
[28] Harrington D A and Van Den Driessche P 2011 Mechanism and equivalent circuits in electrochemical impedance spectroscopy *Electrochim. Acta.* **56** 8005

Chapter 4

Supercapacitor concepts

4.1 Overview

There are indirect and direct ways to store electrochemical energy. In batteries, chemical energy is indirectly stored as charge flows between the electrodes in the faradic oxidation and reduction processes. A direct means of storing electrostatic charges is on the electrodes forming the plates of a capacitor, which is known as a nonfaradic process. The efficiencies of these two processes are not limited by the thermodynamic Carnot cycle, and they involve reversible processes. Furthermore, two types of capacitive behaviors have been described. The first is produced by a double layer at the electrode interfaces, while the other develops during certain electrode processes where the charge (q) passing through the double layer is, to some extent, a function of potential (V). This leads to a derivative, dq/dV, equivalent to a capacitance, which is known as pseudocapacitance, C_{φ}.

Electrochemical capacitors utilize double-layer capacitance at two electrode/solution interfaces, one of which is charged positively and the other of which is charged negatively in an electrolyte solution separated by a separator. Charge–discharge potential of an ideal capacitor is shown in figure 4.1(a).

A counter electrode or reference electrode is used to measure the capacitive behavior of a device with respect to which potential measurements can be controlled and performed. The electrochemical behavior of each electrode in a pair of electrodes can be examined using a reference electrode. Two interfacial drops in potential occur in a capacitor cell, one across each double layer. Also, upon discharge, there is an ohmic drop within the solution, as shown in figure 4.1(b).

The determination of double-layer capacitance and its potential dependence via cyclic voltammetry (CV) [2] and AC impedance provides differential information about the behavior and structure of the interfaces upon charging and discharging. The amount of charge, q, on each side of the electrode interface—corresponding to electron density on the metal side and ion density on the solution side—depends on the measured electrode potential, E (a potential equivalent to the Fermi level of the

doi:10.1088/978-0-7503-5944-3ch4

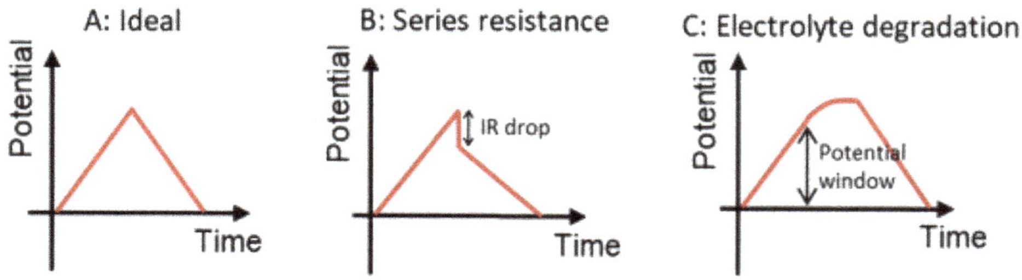

Figure 4.1. The double layer in an electrochemical capacitor at each of two electrodes: (a) charged capacitor at open circuit, (b) capacitor current passing on discharge with an IR drop, (c) electrolyte degradation is shown in by a plateau in potential. Reprinted from [1], Copyright (2022), with permission from Elsevier.

electron energy state in a metal). This is directly determined from the differential capacitance of the double layer, C_{dl}, given by dq/dE. The double-layer capacitance depends on the interface condition between the electrode metal surface and the solvated ions in the solution. The region near the metal surface and the metal interface is assumed to be 'ideally polarizable.' In an ideally polarizable electrode, the changes in potential are due to flow of charge between the electrodes. These only change the charge density on the metal and the ion density on the solution side of the electrode surface, leading to charging of the formed double layer. The change in potential helps charge to flow from the external circuit to the solution, even though no charge passes through the double layer (faradic interaction). Thus, the charged interfaces remain at electrostatic equilibrium at a given potential as opposed to behaving faradically at thermodynamic equilibrium and Au, NaF, NaOH, and Na_2SO_4 provide nearly ideal polarizable interfaces. The total current density is the sum of a double-layer charging component, i_{dl}, and a faradic solution decomposition component, i_F, $i = i_{dl} + i_F$, as discussed earlier in chapter 1.

Thus, the interface at an ideally polarizable electrode is purely a double layer, and AC evaluation of the capacitance of an ideal electrochemical capacitor device shows that it is frequency independent. However, frequency dependence is commonly observed, where the phase angle for double-layer capacitance does not have an ideal value of 90° at all frequencies (the possibility of DC leakage can also occur). Deviation from ideal capacitive behavior may occur at very high frequencies due to dielectric loss associated with solvent orientation polarization in the double layer. Low-frequency losses occur when there are slow anionic chemisorption processes. Such cases of incomplete polarization at the electrode surface cause energy dissipation in the charging and discharging cycles. In a porous electrode, the nonuniform distribution of double-layer capacitance also results in nonideal behavior. Thus, an electrostatic equilibrium is presented by the double layer of an ideally polarizable electrode, while nonpolarizable interfaces induce free charge transfer across the electrode interface, as governed by the electrochemical thermodynamic equilibrium of the Nernstian kind. As per our earlier discussion in chapter 1, the properties of double layer can be described in four sections: (1) the diffuse part of the double layer nearer the solution and away from the innermost contact layer of the ions is known as the Helmholtz layer. (2) The Helmholtz layer is the region where ions are adsorbed. The finite sizes of the ions in the electrolyte determine the distances of closest approach to the surface of the electrode. The Helmholtz layer consists of two planes, namely, the inner Helmholtz plane and the outer Helmholtz plane; the first relates to the anions at the distance of closest approach to the metal electrode, while the second relates to hydrated cations located at longer distances determined by the radii of solvation (or hydration) [3]. (3) The action of adsorbed solvent dipoles and their orientation (which depends on the potential) in the two regions of the double layer creates a third region. (4) A potential-dependent spread of electron density occurs from the metal surface toward the solution. The inner-layer capacitance cannot be interpreted due to varying values at the positive and negative sides of the potential of zero charge (PZC). Different ions exhibit varying interactions due to their different ionic radii. For example, anions tend to be strongly chemisorbed, as their lone-pair donor strength is larger than that of isoelectronic cations. Moreover, the adsorption of large organic cations in nonaqueous solution capacitors has major effects on the interfacial capacitances of the electrodes. The adsorption of anions at the electrodes involves partial charge transfer. The extent of the charge transfer, called 'electrosorption valence,' depends on the polarizability of the anion, the electronic structure of the metal electrode, its donor strength, and the energy of solvation [4]. Nonaqueous electrolytes are desirable due to their higher operating potentials. Moreover, the solvation energies of ions in nonaqueous solvents are less than in water; thus, anions are more strongly chemisorbed onto electrodes when aprotic media are used instead of water.

Solvents are mainly categorized into three types, namely water, heavy water, and methanol and other alcohols. Hydrogen fluoride and carboxylic acids interact, and their molecules orient through hydrogen bonding. Polar solvents have dipole–dipole interactions without hydrogen bonding and molecular directionality; these include dimethyl sulfoxide, propylene carbonate, acetone, esters, and methyl dichloride. The nonpolar solvents include hydrocarbons and fluorocarbons. Moreover, non-aqueous aprotic solvents for electrolytes are attractive, owing to their higher solution decomposition voltages. Hence, they can provide higher energy and power densities.

The Helmholtz layer capacitance, C_H, is the reciprocal of C_{dl} value as $(1/C_{dl})$ at a given potential, calculated by subtracting the reciprocal of diffuse layer capacitance (in dilute solutions). The metal contribution to the double-layer capacitance appears as a series capacitance, C_M, whose reciprocal $(1/C_M)$ is added to the other contributions. At the electrode surface, the electron density varies at three potentials. At a negative potential with respect to the PZC, one may observe a greater penetration of electron spillover on the solution side and adsorbed ions in the Helmholtz layer. This has an impact on the chemisorption of ions (anions) and the orientation and distribution of solvent molecules. For example, protons of oriented H_2O molecules are shielded on the negative side due to electron spillover at the metal surface.

The high surface area of 1000–2000 $m^2\ g^{-1}$ usually found in carbon leads to its use in double-layer electrochemical capacitors, where the diffuse region becomes dominant because the electrolyte extends over a very short distance near the surface of the pores. The diffuse layer at the electrodes extends about 100 nm into the solution for a 0.001 M solution, about 10 nm for a 0.01 M solution, and about 1 nm for a 0.1 M solution of a univalent electrolyte. Moreover, the distribution of the potential is more complex in microporous structures than at infinite planar electrodes. In very fine pores of an electrode (2–10 nm in diameter), the diffuse layer of the double layer in a dilute solution can overlap with the diffuse layer on the other side of the pore. However, for more acidic or alkaline electrolytes, the overlap of the diffuse layer does not occur because the net charge density on the surface of the pores is fully screened by the counter ions in the Helmholtz layer. The details of the impacts of various parameters on energy storage performance will be discussed in a different section of this chapter. We first need to understand the difference between the charge storage mechanism of electric double-layer capacitance and that of pseudocapacitance.

4.2 Electric double-layer capacitance and pseudocapacitance

Electric double-layer capacitance (EDLC) and pseudocapacitance are two dominant mechanisms in energy storage [2]. As mentioned earlier, in electric double-layer capacitors, energy is stored via the electrostatic formation of a charged double layer at the electrode–solution interface. In pseudocapacitors, energy is stored by redox reversible reactions at or near the surface of the electrode, possibly accompanied by electrosorption. Pseudocapacitor materials are usually conductive polymers or metal oxides. Both mechanisms may coexist in electrochemical capacitors [5].

4.2.1 Charge storage mechanism in electric double-layer capacitors

Energy is stored in electric double-layer capacitors via the arrangement of charges in the compact layer (Helmholtz layer) that occurs during the charge–discharge processes without any faradic reaction. This results in a displacement current that can be measured and controlled by various methods. Due to the surface-limited mechanism, this method stores less energy than batteries or pseudocapacitors. The electric double layer (EDL) capacitance is constant, and the response current is given by

$$I = dQ/dt = C_{dl} dV/dt,$$

where t is the charging time and V is the applied voltage, which varies linearly as $V = V_0 + vt$. If V_0 is the initial potential and v is the sweep rate (in V s^{-1}), then the current is given by $I = C_{dl}v$; thus, CV measurements yield a rectangular current (I)–voltage (V) response for various scan rates, figure 4.2. Moreover, the voltage response under charging and discharging at constant current rises and decays at a constant rate in a triangular curve. Many efforts made to improve the constrained energy density in electric double-layer capacitors include matching pore sizes with the electrolyte ions in carbon, doping or modifying carbon with heteroatoms, and designing new electrolytes with a larger working potential over a wider temperature range. The other physical parameter changes considered for graphene, activated carbon (AC), and carbon nanotubes (CNTs) are the number of layers, pore structures, and electrode surface area [5].

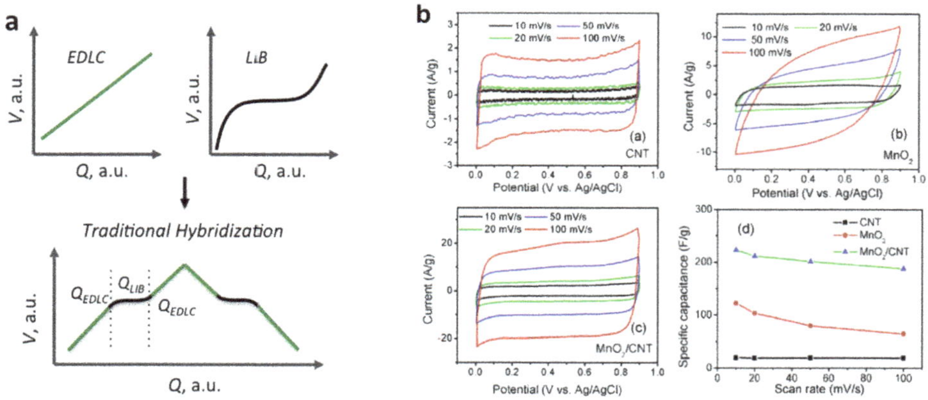

Figure 4.2. (a) Illustration of electrochemical energy storage mechanisms. Adapted with permission from [6]. Copyright (2020) American Chemical Society. (b) CV results for carbon nanotubes (CNTs), pure MnO_2, and MnO_2/CNT nanocomposite electrodes at different scan rates from 10 to 100 mV s^{-1}. (lower right) Specific capacitance vs. scan rate of different electrodes. Reproduced from [7]. CC BY 4.0.

4.2.2 Pesudocapacitance

Among the processes responsible for electrochemical energy storage in solid-state electrodes, pseudocapacitance arises due to faradic redox reactions at the electrode surface, the presence of oxide films (such as RuO_2, IrO_2, and Co_3O_4), or electrosorption processes. The Greek word ‘ψευδής/pseudes,’ which translates as ‘false’ or ‘lying,’ is the origin of the prefix ‘pseudo.’ The faradic origin of pseudocapacitance involves the passage of current across the double layer, while the origin of capacitance depends on the extent of charge on the electrode surface and the corresponding change in potential; thus, the derivative of charge and potential is equivalent to capacitance. Hence, this phenomenon, which differs from electrostatic double-layer capacitance in its origin, is referred to as pseudocapacitance and can be ascribed to highly reversible charge transfer reactions at the surface or intercalation, as shown in figure 4.2 [2].

Perhaps 1%–5% of carbon-based double-layer capacitors have a contribution from pseudocapacitance due to surface reactivity, while batteries (pseudocapacitive devices) have a 5%–10% component from electric double-layer capacitance. Thermodynamically, pseudocapacitance arises when any property, y, corresponding to the passage of charge through the double layer, is related to potential by $y/(1-y) = \mathrm{K}\exp(VF/RT)$; here, y can be the fractional coverage, θ, of an electrode surface; fractional adsorption, X, into an intercalation host; or conversion (of an oxidized species to a reduced species) in a redox system in solution or a hydrous oxide. Thus, pseudocapacitance is the result of a faradic electron transfer process between oxidized and reduced species (through chemical changes in the reactant species, unlike the accumulation or deficiency of electrostatic charges at the double layer of an electrode). In the case of pseudocapacitance, the extent of the reaction, Q, is a continuous function of potential V, so that the derivative dQ/dV has the property of capacitance [4]. Thus, y can be differentiated with respect to V as

$dy/dV = F/RT \cdot K \exp(VF/RT)/[1 + K \exp(VF/RT)]^2$, where dy/dV is proportional to the capacitance. The capacitance reaches its maximum value at the half-conversion of the redox reactants, or half-occupancy of the electrode surface or the bulk materials. Thus, the capacitance of pseudocapacitive systems is 10–100 times larger than the capacitance of double-layer systems of the same mass or volume. The complex behavior of pseudocapacitance can be modeled by including interacting particles (exp $\pm gy$). This additional term lowers the capacitance (corresponding to dy/dV) compared to the case where $g = 0$, but permits a wide range of potential. Moreover, the capacitive component of Warburg impedance is another type of pseudocapacitance, which originates due to electrochemical reactions involving diffusion and varies as the square root of frequency under AC modulation.

In the electrosorption process, pseudocapacitance (C_φ) becomes proportional to $\pm d\theta/dV$. The kinetic equations for two-dimensional surface processes consider the fractional surface coverage, θ, the fraction of unoccupied surface sites $(1 - \theta)$, the effect of potential on the rate of charge transfer in the formation or removal of a two-dimensional layer of the electrochemisorbed species, and the g factor (lateral interaction effects).

Underpotential deposition (UPD) of hydrogen is observed at a positive potential relative to the reversible hydrogen electrode potential. This distinguishes UPD from the process of H deposition that takes place during H_2 formation, which requires cathodic overpotentials. A similar UPD also occurs during the electrodeposition of base metals, such as Cu, Bi, or Pd, on noble metals, such as Au, Ag, or Pt. The condition required for the UPD of two-dimensional monolayers before the formation of the bulk phase is that the Gibbs energy of H-to-M (noble metal substrate) bonds or base metal-to-M bonds must be lower than that of H–H bonds or that of the intermetallic bonds in the bulk phase of the metal finally deposited. The adsorption pseudocapacitance can originate if the kinetics of the processes are sufficiently easy. The Langmuir-type electrosorption isotherm describes pseudocapacitance as follows:

$$C_\phi = \frac{q_1 F}{RT} \times \theta(1 - \theta).$$

However, it does not consider the lateral interaction energy, $g\theta$, which accounts for the repulsion between electronic orbitals during chemisorption in adsorbate-to-metal polar bonding and the electronic effects due to a change in localized surface charge (electron) distribution. Thus, the electrosorption isotherm becomes similar to a Frumkin-type isotherm:

$$\frac{\theta}{1 - \theta} = K \exp(-g\theta) \times c_{H^+} \exp(VF/RT),$$

where $K\exp(-g\theta)$ is an equilibrium constant. K for $\theta = 0$ (Langmuir isotherm) is negligible, while θ increases for positive values of g. C_φ can be obtained as follows:

$$C_\varphi = \frac{q_1 F}{RT} \times \frac{\theta(1 - \theta)}{1 + g\theta(1 - \theta)}.$$

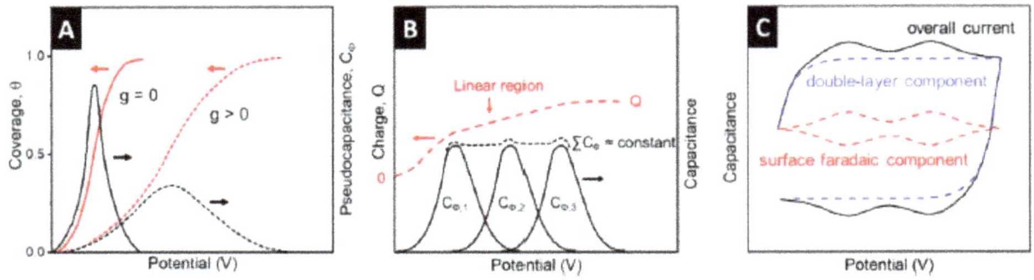

Figure 4.3. (a) The variation of pseudocapacitance, C_{Φ}, and surface coverage, θ, for ideal Langmuir electrosorption ($g = 0$, solid curves) and Frumkin-type electrosorption that involves lateral interactions between surface adsorbates ($g > 0$, dotted curves). (b) and (c) Illustrations of how three redox processes ($C_{\Phi,1}$, $C_{\Phi,2}$, and $C_{\Phi,3}$) can overlap to form an almost-constant overall capacitance over an appreciable potential range in electrochemically formed RuO_2 thin films. Adapted with permission from [8]. Copyright (2020) American Chemical Society.

The maximum value of C_{φ} is lower than the Langmuir electrosorption for positive values of g. In addition, the values of C_{φ} cover a wide range of potential V compared to the Langmuir case are shown in figure 4.3.

On the other hand, C_{φ} vs. V profiles for negative values of g are sharper than in the Langmuir case, with a singularity at $g = -4$, which corresponds to a two-dimensional phase transition. Negative g corresponds to attraction between particles in the formation of a monolayer, and at $g < -4$, an island of two-dimensional structure forms. Further, with increasing positive g values, the electrosorption process becomes difficult due to the increase in potential as the coverage increases. However, electrosorption becomes easier with negative g values as the coverage increases due to attractive forces in the sublayers. Thus, the effective potential required for the same coverage (or the required charge) is less. The charge passing through the layers can be evaluated from the integral under the voltammogram at a sweep rate of $s = dV/dt$ for a sweep between potential limits V_2 and V_1 as follows:

$$\int_{V_1}^{V_2} i(V) \cdot dt \equiv \int_{V_1}^{V_2} i(V) \cdot dV/s,$$

where $i(V)/s$ is the pseudocapacitance C_{φ}, and the term in the integral represents charging from potential V_1 to V_2 and the passage of the charge C_{φ} $(V_2 - V_1)$ through the two-dimensional layer, where C_{φ} varies with the changing potential. Hence, $C_{\varphi} = i/(dV/dt)$, where i is the response current density in response to a potential varying linearly over time in a system with pseudocapacitance. In polycrystalline surfaces, C_{φ} vs. V profiles may show several maxima far from a constant value of C_{φ}. In CV, there is always a rising current (charging current) due to the presence of double-layer capacitance, which may account for 5%–10% of the pseudocapacitance charging current for an appreciable value of C_{φ}. The two can be distinguished by their different frequency dependences. Usually, in the range of 10^5–10^6 Hz, double-layer capacitance is maintained; in contrast, due to electrode kinetic rate limitations, the range for pseudocapacitance is restricted to 10^3–10^4 Hz.

Moreover, after differentiation, the Nernst-type redox equation gives a potential-dependent pseudocapacitance that has the same relationship as the equation

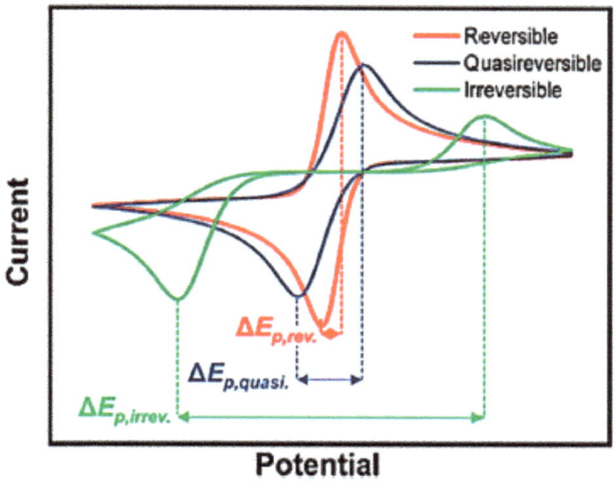

Figure 4.4. Reversible, irreversible, and quasi-reversible cyclic voltammograms. Adapted with permission from [9]. Copyright (2021) American Chemical Society.

representing the electrochemical Langmuir isotherm. For example, Conway *et al* discovered that ferrous disulfide (FeS_2), a salt containing disulfide ions, $^{-}S–S^{-}$, exhibits redox pseudocapacitance over a potential range of 0.8 V. This salt can be reduced via the transfer of a charge of 2e at the surface of FeS_2 to produce two sulfide ions that remain at the interface of the lattice. In a reversible process, the system can be cycled thousands of times without any significant degradation, as shown in the cyclic voltammogram. For comparison, irreversibility in the reduction of FeS (the salt of S^{2-} ions) is observed in the cyclic voltammogram. Figure 4.4 schematically shows the differences between irreversible, reversible, and quasi-reversible cyclic voltammograms [4].

In this process, reversibility arises because there is no phase change, and the interaction is limited to the surface only. The relationships between peak current and voltage vs. potential in pseudocapacitance are given by $di/dt = 0$ and $d^2\theta/dt^2 = 0$, respectively, where $\theta = \frac{K_1 c_{H^+} \exp\left(\frac{VF}{RT}\right)}{1 + K_1 c_{H^+} \exp\left(\frac{VF}{RT}\right)}$, which is the Langmuir type of electrosorption isotherm. The net faradic current density is $q_1 d\theta/dt$, where $\frac{d\theta}{dt} = \frac{K_1 c_{H^+}}{K_1 c_{H^+} + \exp-\left(\frac{V(t)F}{RT}\right)} \times \frac{F}{RT} \times \exp\left(\frac{V(t)F}{RT}\right) \times s$ and s denotes sweep rate such that $V(t) = V(t = 0) + st$. The net faradic current density is $\frac{i}{F} = \frac{q_1}{RT} \times \frac{K_1 c_{H^+} \exp-\left(\frac{V(t)F}{RT}\right)}{\{K_1 c_{H^+} + \exp-\left(\frac{V(t)F}{RT}\right)\}^2}$. s.

The peak potential and corresponding current can be evaluated using $di/dt = 0$ and $d^2\theta/dt^2 = 0$, respectively, as follows:

$$\frac{d^2\theta}{dt^2} = K_1 c_{H^+}\left(\frac{V(t)F}{RT}\right)^2 \exp-\left(\frac{V(t)F}{RT}\right) \times \left[\frac{K_1 c_{H^+} - \exp-\left(\frac{V(t)F}{RT}\right)}{\left\{K_1 c_{H^+} + \exp-\left(\frac{V(t)F}{RT}\right)\right\}^3}\right]$$

The peak potential at a given current density is $V_p = -\left(\frac{RT}{F}\right)\ln K_1 c_{H^+}$, and the peak current density is $i_p = (q_1 F/4RT) \cdot s$, where $q_1 F/4RT$ is the maximum value of C_φ, for $\theta = 1 - \theta = 0.5$ ($g = 0$). The expression for the irreversible deposition of electroactive species can be found in chapter 10 of Conway's book [10].

The faradic process results in a continuous increase in current; however, in the electrochemical adsorption (UPD) or surface redox process, at low potentials, the response current is initially small during a potential sweep; as the potential increases, the current starts to increase exponentially and then drops to zero as the value of $1 - \theta$ eventually becomes zero. In the reversible case, the current reaches a maximum when $\theta = 1 - \theta = 0.5$. This effect is similar to the process of a surface redox reaction, where the active materials in the redox process keep changing as the potential is swept.

The effect of interactions due to $g>0$ and $g<0$ on the shape of the voltammetry curves was evaluated further by Conway's group. They performed sweep reversals by varying the potential at points corresponding to $\theta<1$ (incomplete coverage) and analyzing the resulting shapes. There is wide variation in the potential range of pseudocapacitances. For instance, at $g = 0$, the values of C_φ arise in a narrow potential range of 120 mV, which is extended to 0.5–1.0 V for larger values of g. However, as mentioned earlier, the values of maximum C_φ are reduced significantly from $g = 0$ but remain substantially larger than the double-layer capacitance obtained using the same electrode area.

Materials that show pseudocapacitive behavior exhibit the characteristics of both double layers and batteries. In particular, pseudocapacitive materials exhibit features that are not limited by solid-state diffusion and provide highly reversible electrochemical responses, resulting in high efficiency at high current densities. Certain pseudocapacitive materials do not show faradic peaks; this phenomenon was described by Costentin and Savéant, who combined both double-layer and faradic interactions [11]. They observed that the double-layer charging current was equal to or greater than the faradic current contribution. Compared to faradic processes in batteries, pseudocapacitive processes are highly reversible over extended time periods; they therefore result in mirrored behavior like that seen in cyclic voltammograms, while battery processes involve phase transformations such as conversion and alloying.

With the growth in new materials showing pseudocapacitive behaviors, deeper insights have been developed. Conway's formalisms discussed above were not based on an atomistic description; thus, more effort was needed to understand the variety of materials and different electrochemical processes. In nanomaterials, the concept of intrinsic pseudocapacitance has been presented (figure 4.5), where intercalation reactions do not undergo significant mass transfer and the diffusion length, l, is much smaller than the square root of the product of the diffusion time (t_d) and the diffusion coefficient (D): $l \ll \sqrt{Dt_d}$. Moreover, in nanostructured materials, large numbers of redox sites are located at the interface between the electrode and electrolyte.

Another instance of nearly rectangular cyclic voltammograms of pseudocapacitive materials has been observed for materials such as MnO_2 and RuO_2, which

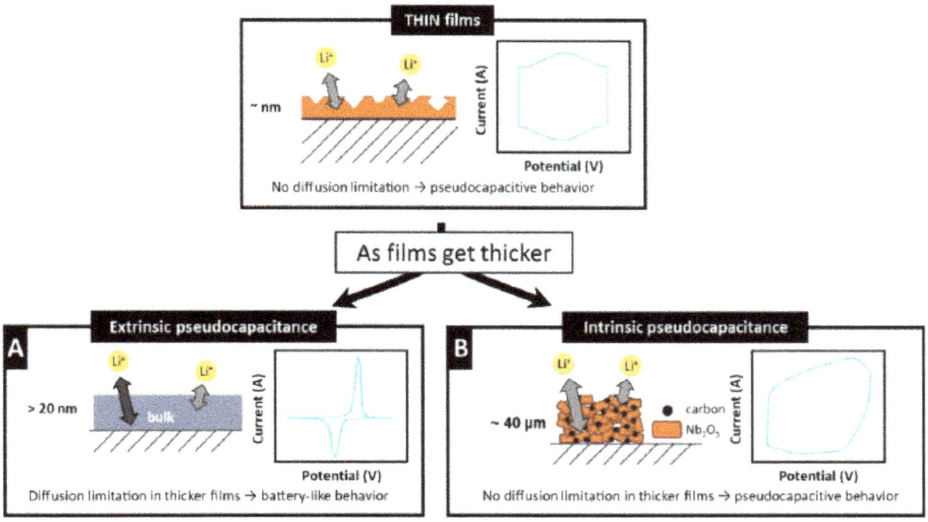

Figure 4.5. Schematic representation of extrinsic and intrinsic pseudocapacitive materials and the corresponding cyclic voltammograms. Adapted with permission from [8]. Copyright (2020) American Chemical Society.

Conway attributed to a series of overlapping redox reactions. However, Costentin and Savéant explained that it was due to the formation of a double layer at metallic conductive electrodes or the contributions of both a surface faradic process and double-layer formation to the overall pseudocapacitance. Similarly, the nomenclatures 'intercalation capacitance' and 'intercalation pseudocapacitance' are analogous. Solid-solution intercalation, where intercalation does not cause a phase transition in the host lattice due to negligible structural transformation, is characteristic of pseudocapacitive processes. In this respect, it is distinct from battery-type processes.

As discussed earlier, the electrochemical characterizations that function as signatures of pseudocapacitance are: (1) a linear relationship between the applied potential and the state of charge (dq/dV), (2) electrochemical near-reversibility, and (3) surface-controlled kinetics without significant transformations (such as phase transformations or the mass transfer of heavy atoms). The electrochemical behavior of pseudocapacitive materials should be measured for durations of less than 10 minutes to observe higher capacities and efficiency than those of typical batteries, as shown in figure 4.6, which depicts the cyclic voltammograms and charge–discharge behaviors of various pseudocapacitive materials.

The cyclic voltammograms of pseudocapacitive materials show broad distributed peaks corresponding to charge transfer, which mirror each other when scanned between both the cathode and the anode. In pseudocapacitive materials, the peak potential as a function of sweep rate does not show a change in the peak position at different sweep rates; it thus provides an important distinction from battery-type materials. Moreover, the energy storage efficiency of such materials is larger for short charge/discharge times (<10 minutes). At high scan rates, large ohmic losses may lead to peak separation in pseudocapacitive materials.

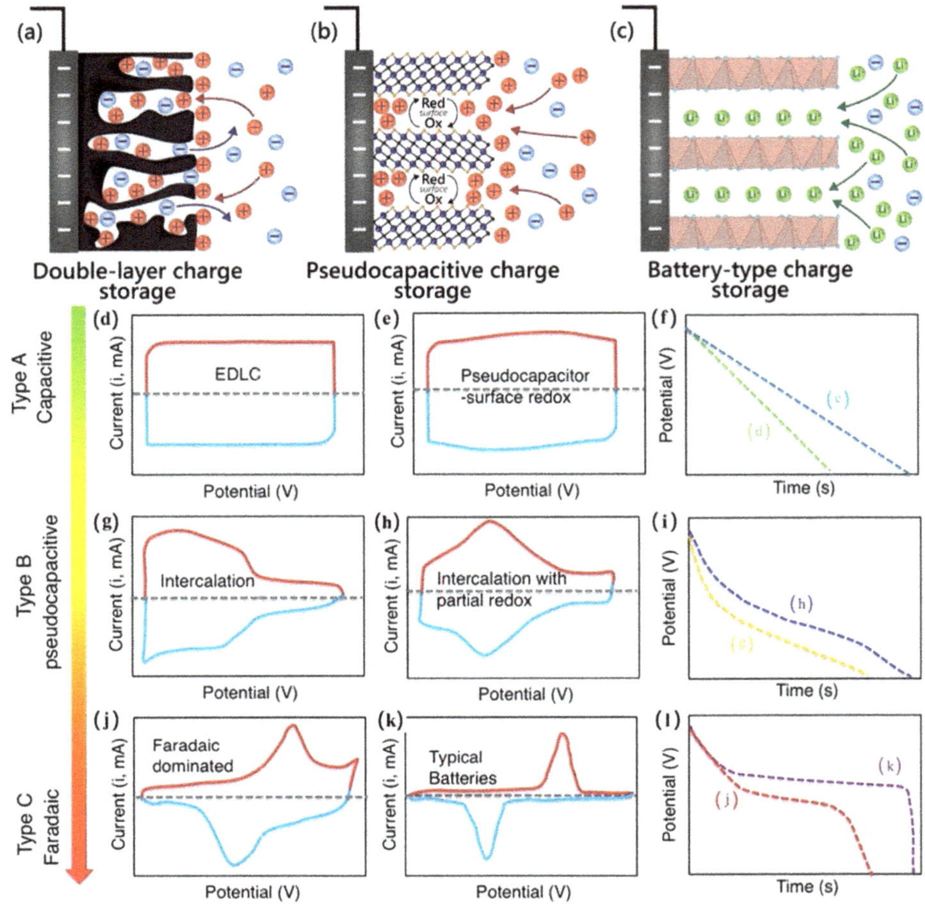

Figure 4.6. (a)–(c) Schematics showing different types of pseudocapacitance, namely surface redox pseudocapacitance, intercalation pseudocapacitance, and the combination of intercalation and surface redox pseudocapacitance. (d)–(l) Corresponding galvanic charge–discharge profiles. Parts (a)–(c) [12] John Wiley & Sons. © 2019 WILEY-VCH Verlag GmbH & Co. KGaA, Weinheim. Parts (d)–(l) adapted with permission from [8]. Copyright (2020) American Chemical Society.

The relationship between the applied sweep rate and the resulting electrochemical current can be used to determine surface-controlled capacitive or semi-infinite diffusion-controlled kinetics as follows: $i(V) = av^b$, where $i(V)$ is the current at a potential, a and b are constant parameters, and b can be determined from the slope of $\log(i)$ vs. $\log(v)$ for various sweep rates. The two limiting cases are considered for $b = 0.5$ and $b = 1$. For $b = 1$, the current is directly proportional to the sweep rate, and the process is surface controlled, which includes the formation of an electrical double layer. In this case, $i = C_{dl}Av$, where C_{dl} is the double-layer capacitance per surface area and A is the electrochemically active surface area of the electrode. The current due to redox processes not limited by diffusion is as follows: $i = \frac{nF^2}{4RT}A\Gamma^*\nu$, where n is the number of transferred electrons, and Γ^* is the amount of redox-active adsorbed species at the surface (mol cm^{-2}). For $b = 0.5$, the reaction is semi-infinite diffusion controlled, and the current is given by the Randles–Sevcik equation, as follows: $i = 0.4958n \cdot FAC^{*}D^{1/2}\left(\frac{\alpha nF}{RT}\right)^{1/2}\nu^{1/2}$, where C^{*} is the maximum concentration of the reduced species in the structure, and α is the transfer coefficient. For b values varying between 0.5 and 1.0, the process is either mixed control (a linear combination of diffusion and capacitive contributions) as follows: $i(V, \nu) = k_1(V)\nu + k_2(V)\nu^{1/2}$ or finite-length diffusion (where the diffusion distance is smaller than $\sqrt{Dt_d}$). Dunn *et al* used the mixed control concept to deconvolve the capacitive vs. diffusive contributions to the total current for several nanostructured metal oxides by rearranging, as $\frac{i(V,\nu)}{\nu^{\frac{1}{2}}} = k_1(V)\nu^{\frac{1}{2}} + k_2(V)$. The values of k_1 and k_2 at specific potentials and different sweep rates provide the quantitative relation between the currents due to surface and diffusion-controlled processes. This method separates cyclic voltammograms into surface-controlled and diffusion-controlled regions. The charges stored due to surface- and diffusion-controlled current contributions can be quantified at different sweep rates.

As mentioned earlier, the ohmic losses and polarization increases with increasing sweep rate and thus shift the redox peak potential; however, the above formulation does not account for this variation but provides a simplistic approach to quantify how nanomaterials develop pseudocapacitive properties.

An alternate method used to identify the charge storage mechanism is to analyze the electrochemical impedance. The impedance data is modeled to produce an equivalent circuit model, which allows one to interpret the electrochemical behavior by distinguishing pseudocapacitance from double-layer capacitance. Figure 4.7 shows the various modeling elements; R_{el}, R_F, CPE_{dl}, and CPE_{pseudo} are the resistance of the electrolyte, the faradic charge transfer resistance, a constant phase element for double-layer capacitance, and a CPE for pseudocapacitance, respectively. The impedance of porous pseudocapacitive electrodes is represented by a CPE for the distribution of capacitance values given by $Z_{\text{CPE}} = [B(j\omega)^p]^{-1}$, where j is an imaginary unit, ω is angular frequency, and B and p are frequency-independent adjustable constants. Here, $p = 1.0$ represents ideal capacitive behavior, and B is the capacitance of the electrode. For $p = 0.5$, Z_{CPE} is the Warburg diffusion impedance of a semi-infinite medium.

Two-dimensional Bode plots provide useful information about pseudocapacitance, where the dependent variables are capacitance C, the phase angle φ, and the angular frequency ω. The capacitance is measured as the response of an electrode to an applied perturbation in such a way that its impedance can be modeled as a frequency-dependent resistor and capacitor in series in an equivalent circuit. Both the imaginary (C'') and real (C') components of the capacitance can be obtained, as follows: $C'(\omega) = \frac{-z''}{\omega\,|\,Z\,|^2}$ and $C''(\omega) = \frac{-z'}{\omega\,|\,Z\,|^2}$. The real component of the capacitance, C', measures the reversible charge stored at a given frequency; at low frequencies, it also corresponds to the capacitance of the electrode measured in cyclic voltammograms. The imaginary capacitance, C'', relates to the energy dissipated in a kinetically irreversible process or under diffusion limitations. The maximum of C''

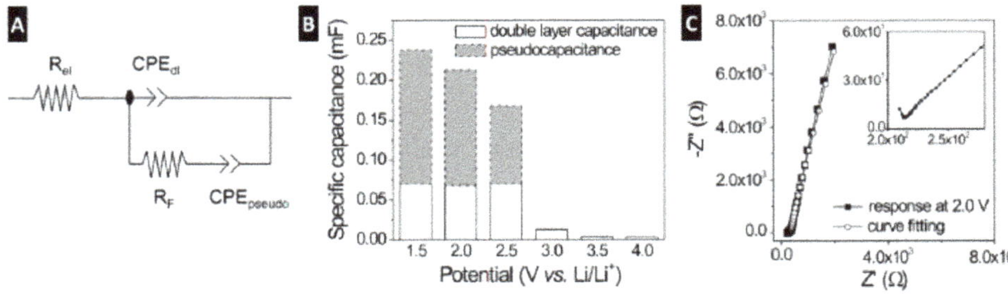

Figure 4.7. Impedance data modeling used to determine capacitive contributions via a k_1, k_2 analysis of CV data. Adapted with permission from [8]. Copyright (2020) American Chemical Society.

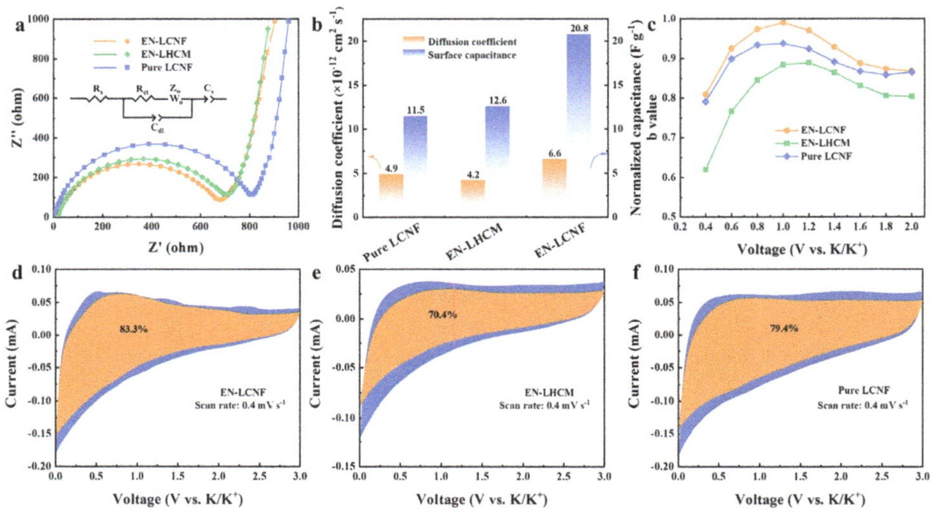

Figure 4.8. Impedance data modeling used to determine capacitive contributions via a k_1, k_2 analysis of CV data. Reproduced from [13]. CC BY 4.0.

vs. frequency is the dielectric relaxation time constant, τ, which varies as the reciprocal of the characteristic frequency. At this point, the circuit transforms from purely resistive to capacitive. The relative contributions of the capacitive and resistive elements in the circuit are given by the phase angle, φ. For ideal capacitors, $\varphi = -90°$; for diffusion-controlled processes, $\varphi = 45°$; and for resistors, it is 0°. Figure 4.8 compares the normalized capacitance versus frequency and the potential for capacitive and pseudocapacitive electrodes. A constant C' value is observed for both capacitive and pseudocapacitive materials over a wide range of potentials at low frequencies, in contrast to the battery-like responses.

4.3 Supercapacitor design, fabrication, operation, and experimental parameters

An electrochemical cell is a vessel containing different components used to conduct CV experiments. The configuration of an assembled coin-cell supercapacitor is

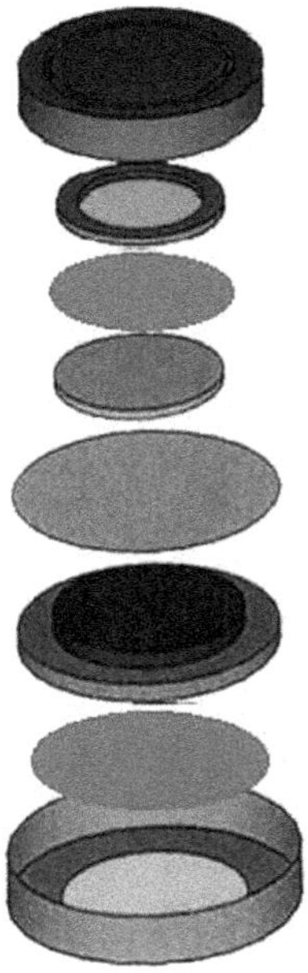

Figure 4.9. Components of a typical energy storage device. Assembly of a coin cell supercapacitor. Reproduced from [14]. CC BY 4.0.

shown in figure 4.9 and a three-electrode setup was shown in figure 3.1. The setup includes a working electrode, a reference electrode, and a counter electrode. The reference electrode is used to accurately measure the potential relative to the control reaction. It has a well-defined, stable equilibrium potential. Common reference electrodes include the saturated calomel electrode (SCE), the standard hydrogen electrode (SHE), and the AgCl/Ag electrode, which are used in aqueous media. The electrochemical interaction of interest occurs at the working electrode. The type of working electrode material varies based on the desired potential window or surface adsorption/reaction. The counter electrode completes the circuit, and the current is measured between the working and counter electrodes. A platinum wire is typically used as the counter electrode. As electrons flow from the electrode to the analyte, ions move in solution to compensate for the altered charge, closing the electrical circuit. The electrolyte solution is commonly a mixture of a solvent and a supporting electrolyte, which is a salt dissolved in the solvent to decrease the solution resistance.

Electrolytes have a solution resistance and an uncompensated resistance between the working and reference electrodes, which is known as an ohmic drop. According to Ohm's law, the potential difference due to the ohmic drop is given by the current passed, i, times the uncompensated resistance, R_u. Large ohmic drops lead to inaccuracies in data measurements. There are three ways to minimize the ohmic drop: (1) reduce i by selecting a slow scan rate or reducing the size of the working electrode; (2) increase the conductivity of the solution by increasing the concentration of the electrolyte, which reduces R_{sol} and thus R_u; (3)

reduce the distance between the two electrodes (working and reference) to directly reduce R_u.

The CV measures the charge transfer between the working and reference electrodes in both redox and capacitive processes. The x-axis in a CV measurement represents the applied potential (E), while the y-axis represents the response current (i) passing through the external circuit. The scan rate (speed) is the parameter related to the linear variation of the potential per second. In a redox process, the equilibrium is described by the Nernst equation, which relates the potential (E) of an electrochemical cell to the standard potential. In CV, this relationship is reflected in the peaks corresponding to cathodic and anodic currents. The dependence of the peak current versus the square root of the scan rate is linear for freely diffusing redox species. This is again represented by the equation, $i_p = 0.446nFAC^0\left(\frac{nF\nu D_0}{RT}\right)^{1/2}$, where n is the number of electrons involved in the electrochemical process, A is the electrode surface area, D_0 is the diffusion coefficient of the analyte, and C^0 is the bulk concentration. The diffusion layer decreases at faster scan rates. The separation between redox peaks is caused by diffusion; thus, the distance (separation) between the two peaks can shed light on the mechanism of charge transfer. As an example, consider an equilibrium between ferrocenium (Fc^+) and ferrocene (Fc). When the potential is scanned during the CV measurement, the Fc^+ reduces to Fc at the negative electrode, producing a current. Duck-shaped voltammograms are obtained due to the dependence of the concentration of Fc^+ versus Fc from the electrode surface on the electrode potential and the way in which ionic species move between the electrode and the solution. As the potential is scanned to a negative potential, the Fc^+ starts depleting/reducing to Fc. The cathodic peak current shows the diffusion of Fc^+ from the solution. The solution near the electrode containing Fc is called the diffusion layer, which grows during the entire scan. Upon scanning to a more negative potential, the rate of diffusion of Fc^+ from the solution decreases, resulting in reduced current. When the direction of the scan is reversed (i.e. scanning in the positive direction), the Fc present at the electrode oxidizes back to Fc^+ at a positive applied potential; thus, there are two peaks corresponding to the maximum cathodic and anodic currents in a typical reversible Nernst equilibrium. In a quasi-reversible process, the separation between peak-to-peak deviates due to diffusion or the surface adsorption of species [5].

We now discuss the CV of an ideal capacitive system at a slow sweep rate. It shows a symmetric response, with no shift or only a small shift in the potential between cathodic and anodic peaks at various scan rates. The deviation from ideal capacitive behavior is due to an effective equivalent series resistance, which causes polarization effects. At increasing sweep rates, kinetic reversibility is lost from the surface faradic process in pseudocapacitors. The peak potential remains independent of the scan rate ν until a critical sweep rate, ν_o, is reached, where the peak potential increases as log ν and the cyclic voltammogram no longer resembles a mirror image after a full scan (as shown in figure 4.10). In pseudocapacitors, kinetic reversibility can be regained at lower scan rates, as no phase change occurs during the charging/discharging process [8].

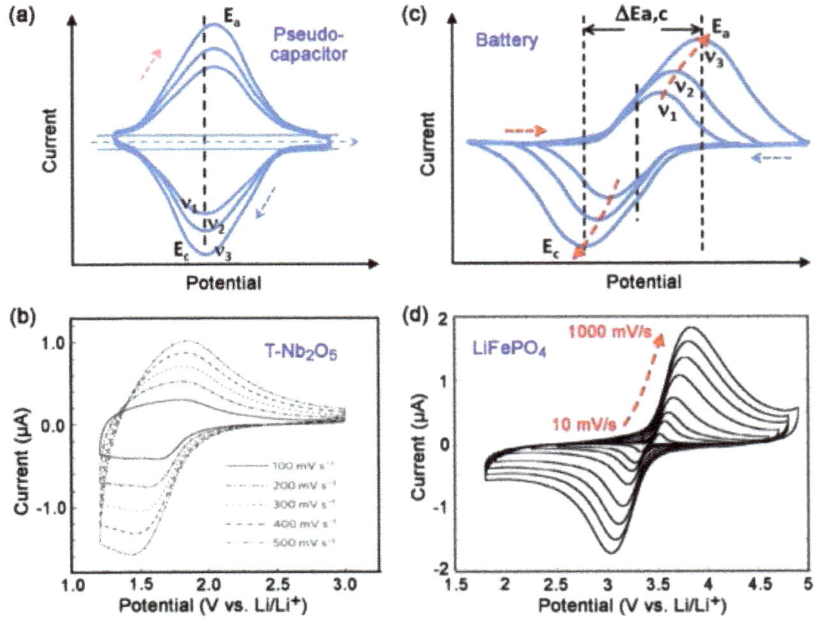

Figure 4.10. CV curves of (a) and (c) are schematic illustrations of pseudocapacitive and batter-type materials, respectively. (b) and (d) are the CV responses from T-Nb_2O_5 as pseudocapacitive material and $LiFePO_4$ to represent the battery-type interaction materials at different sweep rates. Reproduced from [15]. CC BY 4.0.

Therefore, CV measurements of pseudocapacitors show broad redox peaks with no shift or a constant shift over a wide range of sweep rates. On the other hand, the cyclic voltammograms of batteries show a large increase in the shift as the scan rate increases, as they undergo crystallographic phase transformation during the electrochemical processes [16], even at lower scan rates (figure 4.11).

The cyclic voltammogram of an electric double-layer capacitor is rectangular, with little or no dependence on the scan rate (figure 4.12). Pseudocapacitors can have rectangular cyclic voltammograms with or without cathodic/anodic peaks for different materials. Thus, it is difficult to differentiate between these two mechanisms (especially when both EDLC and pseudocapacitance are present in a material). It is important to understand the distinction between capacitive processes and diffusion-controlled intercalation processes when selecting materials and designing electrodes. As described earlier, when the process is controlled by surface-limited interactions/reactions, the response current is linearly dependent on the scan rate, ν ($di/d\nu$=constant=capacitance), while for semi-infinite diffusion, the response current linearly varies with $\nu^{1/2}$ (i.e. $di/d\nu^{1/2}$ is constant). This allows one to quantitatively distinguish the current distributions of the two processes. For MnO_2/Au (core–shell) nanostructure, the surface-dominated capacity is estimated to be 88 F g^{-1} at 5 mV s^{-1}, which remains constant for a wide range of scan rates. The diffusion-controlled insertion capacity is shown to be dependent on the scan rate. Another method used to distinguish the two charge storage mechanisms is to

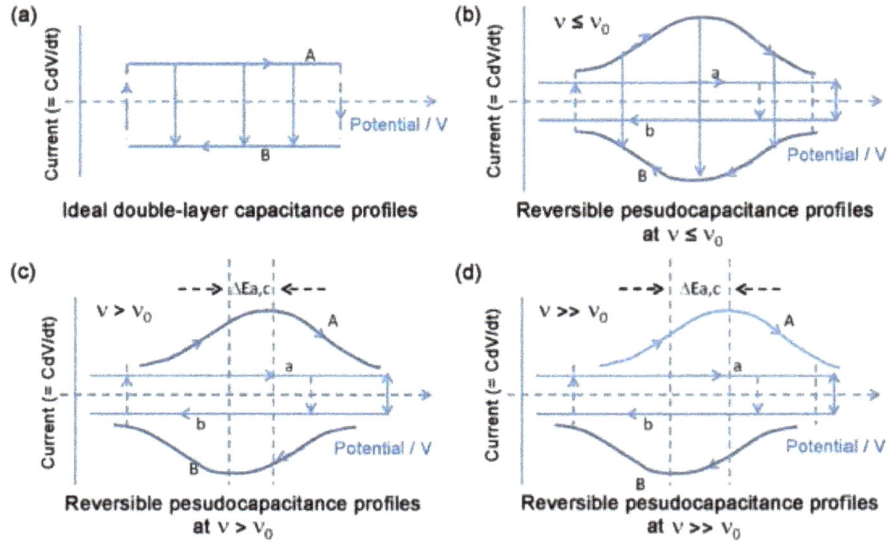

Figure 4.11. CV responses of (a) ideal double-layer capacitor and (b)–(d) reversible pseudocapacitors at different sweep rates v, where v_0 is the critical sweep rate. Reproduced from [15]. CC BY 4.0.

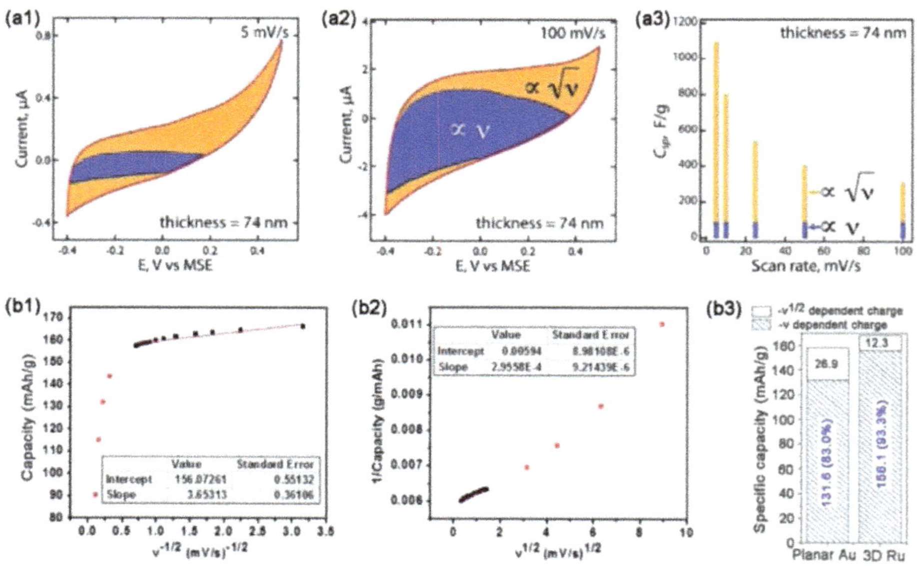

Figure 4.12. Two different methods for deconvolving the surface ($\propto v$) and bulk charges ($v^{1/2}$). Reproduced from [15]. CC BY 4.0.

establish a relation between the stored charge and the scan rate. The total charge, q^*, includes capacitive (q_s^*) and diffusion-controlled (d^*_d) contributions, which scale with the scan rate as ν and $\nu^{1/2}$, respectively. The q_s^* value for the surface-dominated process is obtained at an infinite scan rate ($\nu \rightarrow \infty$), where $q^* \rightarrow q^*_s$, while the maximum charge density (q_m*) is obtained when ($\nu \rightarrow 0$) and $q^* \rightarrow q^*_m$. Therefore, the total charge can be written as $q^*(\nu) = q^*_{\nu \rightarrow \infty} + q^*_d = q^*_{\nu \rightarrow \infty} +$ const$(\nu^{-1/2})^*$, which includes the contributions from the capacitive and diffusion-controlled processes.

4.4 Equivalent series resistance

In an ideal supercapacitor, the real and imaginary impedances should be out of phase by 90° in response to a sinusoidal alternating potential. This phase relationship should be independent of the signal frequency. Practically, capacitors exhibit a phase angle less than 90° and exhibit a dependence on frequency. This behavior is due to a series resistance offered by the capacitor known as the ESR(esr), which mainly originates from the resistance of the electrolyte and/or contact resistances. This can be represented by an equivalent circuit, as shown below, which has a frequency-dependent impedance, $Z(\omega) = esr + 1/j\omega C$. At an infinite ω, $Z(\omega) = esr$.

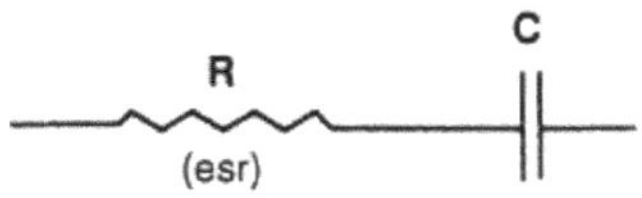

ESR values vary from milliohms to a few tens of milliohms and increase for high-voltage applications based on nonaqueous electrolyte solutions, which have lower specific conductance. If a series capacitor is shorted, all its energy is dissipated with a time constant given by $t_{\text{cutoff}} = R_{\text{esr}} \times C$ (in seconds). At the start of discharge, the product of initial voltage and current, i.e. the power, P, depends on the energy stored in the capacitor, such that $P/E = 2/t_{\text{cutoff}}$ s^{-1}. Thus, the ratio P/E determines the physical construction and materials of the capacitor. The power at the time of short circuit is given by $P = V^2/R_{\text{esr}} = CV^2/t_{\text{cutoff}}$ W, where V is the voltage at the start of discharge and C is the capacitance.

4.5 Leakage resistance

The energy stored in a supercapacitor is directly proportional to the square of the cell potential; hence, it is desirable to increase the cell potential using solvents like nonaqueous and ionic liquids with wide electrode potential windows. However, at extreme cell potentials, the solvent may decompose due to material stability or the thermodynamic window of the solvent, which may result in self-discharge or an increase in cell pressure due to the generation of gases. Other challenges that supercapacitors face are leakage current, rapid discharge, and reduced shelf life.

In a double-layer supercapacitor, no charges cross the double-layer interface (within a certain electrode potential range), and the charging and discharging current density passes through the supercapacitor. However, several unwanted processes occur, accompanied by a faradic leakage current density (i_F) at higher electrode potentials beyond the electrochemical decomposition limits of the electrolyte. Therefore, the total current density required to charge a supercapacitor is as follows:

$$i_{\text{cell}}(\text{charging}) = i_{dl} + i_{lk} + i_{iF} \text{ or } i_{\text{cell}}(\text{discharging}) = i_{dl} - i_{lk} - i_{iF}.$$

This shows that, due to the leakage current density (i_{lk}) and faradic leakage current density (i_{iF}), the charging current of a supercapacitor is larger than expected, and the current obtained upon discharge is smaller than expected. The two leakage current densities cause self-discharge in supercapacitors. The main factors contributing to parallel leakages in supercapacitors are electrolyte impurities (which cause faradic reactions), nonuniformity of charge acceptance on the electrode surface (caused by porosity, for example), and short circuits between the two electrodes. Since both solvent decomposition and parallel leakage occur within the cell during the electrochemical processes, the charging and discharging of the cell can only be described using electrochemical parameters such as the ESR, capacitance, and other electrode constants.

4.6 Supercapacitor charging and discharging

The charging and discharging of a supercapacitor can be done either at a constant cell potential (to record the response of the cell current varying with time) or at a constant current (to record the variation in cell potential over time). In a general scenario, in the absence of both the parallel leakage process and the electrochemical decomposition of the solvent, we define the following: I_{cell} is the constant current during the charging or discharging of the supercapacitor; V_{cell} and C_{dl} are the cell voltage and double-layer capacitance, respectively; V_{sc} is the cell voltage across the double-layer capacitance; and i_{sc} is the current at the time of charging or discharging, which is equal to i_{cell}. When a supercapacitor starts to charge from the zero charge state, the voltage across it is given by

$$V_{sc} = \frac{1}{C_{dl}} \int_{t=0}^{t=t} i_{sc}\, dt = \frac{1}{C_{dl}} \int_{t=0}^{t=t} i_{\text{cell}}\, dt = \frac{I_{\text{cell}}}{C_{dl}} dt = \frac{I_{\text{cell}}}{C_{dl}} t.$$

Then, the cell voltage of the supercapacitor becomes linearly proportional to time, as follows:

$$V_{\text{cell}} = I_{\text{cell}} R_{\text{esr}} + V_{sc} = I_{\text{cell}} R_{\text{esr}} + \frac{\text{I}_{\text{cell}}}{C_{dl}} t \quad \left(\text{charge process}\right).$$

At a maximum design potential of $V_{\text{cell}}{}^{\text{max}}$, the voltage across the double-layer capacitor when discharging begins is

$$V_{sc}^{0} = V_{\text{cell}}^{\text{max}} - I_{\text{cell}} R_{\text{esr}}.$$

Therefore, the cell voltage is given by

$$V_{\text{cell}} = V_{sc}^{0} - I_{\text{cell}} R_{\text{esr}} - \frac{I_{\text{cell}}}{C_{dl}} t \quad \left(\text{discharge process}\right).$$

The maximum energy and power densities can be evaluated using the relations $(E_m)_{\max} = 1/2\ C_m(V_{osc})^2$ and $(P_m)_{\max} = 1/4\ _m(V_{\mathrm{osc}})^2/R_{\mathrm{esr}}$, respectively.

Now, the presence of the voltage-independent leakage process is accounted for by adding an additional equivalent leakage resistance (R_{lk}), which may have a significant impact on the variation in the cell potential. Under these conditions, the supercapacitor charging voltage is given by

$$V_{sc} = I_{\mathrm{cell}} R_{lk}\left(1 - \exp\left(-\frac{t}{R_{lk} C_{dl}}\right)\right) \left(\text{charging process}\right),$$

and the cell voltage is as follows:

$$V_{\mathrm{cell}} = I_{\mathrm{cell}} R_{\mathrm{esr}} + I_{\mathrm{cell}} R_{lk}\left(1 - \exp\left(-\frac{t}{R_{lk} C_{dl}}\right)\right) \left(\text{charging process}\right).$$

This shows that at $t = 0$, $V_{\mathrm{cell}} = I_{\mathrm{cell}}R_{\mathrm{esr}}$, and when $t = \infty$, $V_{\mathrm{cell}} = I_{\mathrm{cell}}R_{\mathrm{esr}} + I_{\mathrm{cell}}R_{lk}$. For $R_{lk} = \infty$, the cell potential becomes

$$V_{\mathrm{cell}=I_{\mathrm{cell}} R_{\mathrm{esr}} + I_{\mathrm{cell}} \frac{t}{C_{dl}}} \left(\text{charge process}\right).$$

After the supercapacitor has been charged to a maximum value, $V_{sc}{}^{0}$, a discharge process at a constant current is described by the following relation:

$$i_{sc} = \frac{(V_{sc}^{0} + I_{\mathrm{cell}} R_{lk})}{R_{lk}} \exp\left(-\frac{t}{R_{lk} C_{dl}}\right),$$

and the voltage across the double layer is

$$\begin{aligned} V_{sc} &= V_{sc}^{o} - \frac{1}{C_{dl}} \int_{t=0}^{t=t} i_{sc}\, dt \\ &= V_{sc}^{o} - \left(V_{sc}^{o} + I_{\mathrm{cell}} R_{lk}\left[1 - \exp\left(-\frac{t}{R_{lk} C_{dl}}\right)\right]\right). \end{aligned}$$

The cell voltage is given by

$$V_{\mathrm{cell}} = -I_{\mathrm{cell}} R_{\mathrm{esr}} + V_{sc}^{0} - \left(V_{sc}^{0} + I_{\mathrm{cell}} R_{lk}\right) \times \left[1 - \exp\left(-\frac{t}{R_{lk} C_{dl}}\right)\left(\text{discharge process}\right)\right].$$

Furthermore, the circuit has constant-current charging in the presence of electrolyte decomposition. Compared to the previous circuit, a parallel electrochemical process is included for potential-dependent current and voltage due to the decomposition of the electrolyte ($V_{sc} = V_f$). It considers electrolyte decomposition at both electrodes using the relations given below:

$$S \rightarrow n_p e^- + P_P$$

$$S + n_n e^- \rightarrow P_P.$$

Here, n_p is the total number of electrons of S oxidation required to produce P_p, and n_n is the total number of electrons of S reduction required to produce P_n. The reactions are irreversible.

4.7 Energy and power densities

In 1968, Ragone presented a plot of specific power density (W kg^{-1}) versus energy density (Wh kg^{-1}) on a logarithmic scale; such plots later became known as Ragone plots. These curves have a hooked shape, corresponding to the decay of energy density, since power density is higher during the discharge of a battery (figure 4.13). The origin of this shape is an increase in electrode polarization and hence reduced cell voltage, because the electrical power density is given by the cell voltage (V) multiplied by the discharge current density, i: *power density* $= iV$. This shows that as i increases, the product iV (and hence the power, P) tends to increase; however, upon a further increase in i, V begins to decay due to polarization losses. In Ragone plots, the initial slopes for supercapacitors (on the log–log scale) are comparable to those of battery systems, which indicates the role of similar iR and i^2R losses caused by internal resistance [9].

In Ragone plots, smaller energy densities are related to higher power densities for any given system. Supercapacitors are capable of delivering high power on discharge, despite exhibiting low energy densities.

In the case of batteries, the theoretical energy density is the product of the faradic charge, Q, that can be withdrawn from a battery at a potential, V, per unit mass of the active material. For a material of mass M, where z is the equivalent in moles, then zFC are the total moles that are available for a complete redox reaction using M. The energy density of a chemical reaction at a constant reversible potential, V, when all the material is consumed, is given by the energy density $= zFV/\mathrm{M}$ J g^{-1} or

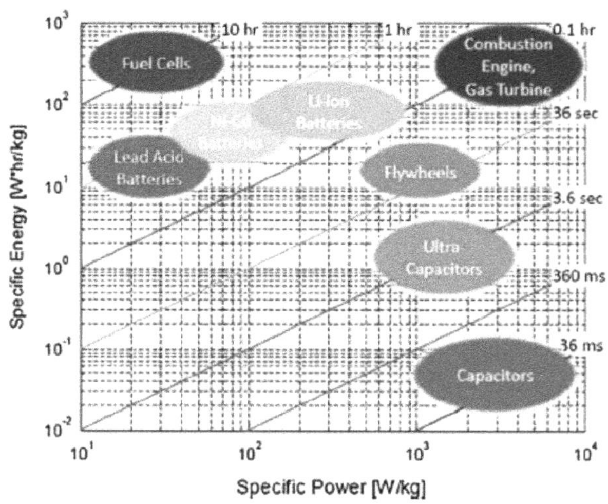

Figure 4.13. Ragone plots for batteries, fuel cells, and other motive-power sources. Reprinted from [17], Copyright (2014), with permission from Elsevier.

$1000zF \cdot V\,M^{-1}$ J kg^{-1}; the energy in Wh is 1 VA × 3600 s = 3600 J. Therefore, the energy density is given by $1000zF \cdot V/3600$ M = $zF \cdot V/36$ M. The density factor in power density is the magnitude of the current delivered per unit area or per unit weight (per gram or kilogram) of the material. In the case of double-layer supercapacitors, the stored electrostatic charge originates from removal or addition, whereas a small fraction of the charge is associated with the faradic process. In contrast, in the case of pseudocapacitors (redox system), the stored electrostatic charge arises from faradic reactions in addition to the 5%–10% that arises from double-layer capacitance. These two processes can be distinguished by their relaxation times or equivalent frequencies. In double-layer supercapacitors, there are no kinetic limitations on power density or the rate of charge or discharge, whereas there may be such limitations for faradic processes in pseudocapacitors (Tafel-type polarization) [18].

By definition, the power delivered by a device through a load resistance corresponds to a certain rate of discharge at a certain current density, as given by *power* = *current* (I) × *voltage* (V). As mentioned earlier with increasing I, V decreases due to polarization losses of voltage, such as an ohmic IR drop due to the internal resistance of the electrolyte and electrodes, in addition to polarization effects associated with Tafel's relation (log(current density) and overvoltage, η) as given by $\ln i = \ln i_0 + \alpha\eta F/RT$, where α is the transfer coefficient of the irreversible kinetic process of the electrode, and i_0 is the exchange current density at zero overvoltage, η. If there are no polarization losses, power increases linearly with I. The power relation, including polarization losses, is: power (P) = $I(\Delta V) = I[\Delta V_{rev} - IR - \Sigma\,\eta(I)] = I\Delta V_{rev} - I^2R - I\Sigma b\ln i + I\Sigma b\ln i_0$, where instances of b denote the Tafel slopes $RT/\alpha F$ of the anode and cathode reactions, $i = I/A$, and $i_0 = I_o/A$ (here, A is the electrode area or mass of the active materials used in the electrode). The negative terms corresponding to ohmic and kinetic polarizations in the equation for P first increase with the current, I, but then decay with increasing I. Hence, the term has a maximum in P, which can be drawn after a certain increase in I. For the maximum power, $dP/dI = \Delta V_{rev} - 2IR - (I\,\Sigma Ab/I + \Sigma b\ln I/A) + \Sigma b\ln I_0/A = \Delta V_{rev} - 2IR - (\Sigma Ab + \Sigma b\ln I/A) + \Sigma b\ln I_0/A = 0$ for $P_{(max)}$. Thus, the rate of decrease of power with increasing current shows the involvement of a complex behavior. The energy density, Q, depends on ΔV and the polarization terms, such that energy density = $Q(\Delta V_{rev} - IR - \Sigma b\ln I/A + \Sigma b\ln I_0/A)$, and d(energy density)/$dI = -QR - \Sigma bQ/I$. At the maximum or minimum energy density, $\Sigma b = -IR$. The ratio, r, of power density ($I \cdot \Delta V$) to energy density ($Q \cdot \Delta V$) can be written as power density/energy density = $r = I \cdot \Delta V/Q \cdot \Delta V = I/Q$. In addition, $Q = I \cdot \Delta t$, Δt is the duration over which a constant current must flow to discharge the capacitor; hence, $r = 1/\Delta t$. Therefore, for a given charge, Q, the shorter the duration of discharge, the greater the power delivered, which is independent of polarization factors.

4.8 Concepts of batteries and supercapacitor energy storage

As mentioned in the previous section, electrical energy can be stored in two different ways: indirectly in batteries and directly as an electrostatic charge. In batteries, electron transfer occurs primarily through faradic redox reactions of chemical reagents, which are converted into electrical energy that flows between the electrodes. Another nonfaradic process causes charges of opposite polarity to form electric double layers, which store charge on electrodes without any chemical modifications, unlike batteries. Batteries store chemical energy through the chemical conversion of the electrode materials (anode and cathode) and phase changes. The key factor in the two mechanisms is reversibility; batteries often exhibit irreversibility of the faradic process upon discharge. Thus, the cycle life of batteries is limited to a few thousand charge–discharge cycles, depending on the battery. However, in supercapacitors, there is no chemical decomposition, and phase changes are not involved during their charging and discharging processes. In supercapacitors, the actual electron charges are either an excess or a deficiency that accumulates on the electrodes, which involves delocalized conduction-band electrons of the metallic electrode. In double-layer charging, there are cases where partial electron transfer occurs, giving rise to pseudocapacitance. In faradic battery-type processes, electrons are transferred to or from the valence bands of the cathode and anode materials.

4.9 Comparison of batteries and supercapacitors

Table 4.1 provides comprehensive details of types of capacitors based on different modes of energy storage (electrostatic or faradic in pseudocapacitance). Batteries are classified as rechargeable (secondary) or nonrechargeable (primary). The materials used in the construction of batteries are also listed in table 4.1.

4.10 Density of charge storage

Let us consider a carbon-based supercapacitor operating at 1 V with an electrode area of 1000 $m^2\ g^{-1}$ and a specific capacitance of 30 $\mu F\ cm^{-2}$. The theoretical stored energy is 150 $kJ\ kg^{-1}$ or 42 $Wh\ kg^{-1}$. On the other hand, if we compare this to the energy storage capability of a Ni–Cd battery at a working voltage of 1.2 V, its ideal energy density is 810 $kJ\ kg^{-1}$ or 225 $Wh\ kg^{-1}$, an increment of nearly 400% in the charge storage of the battery.

4.11 Comparison of charge and discharge behaviors in supercapacitors and batteries

The process used in batteries has a maximum Gibbs energy of $G = Q \cdot \Delta E$, where Q is the charge and ΔE is the potential difference; however, for supercapacitors, it is $G = (1/2)QV$. It is clear that for a given potential difference, $\Delta E = V$, the energy stored in a cell for a faradic charge Q is twice that stored in a supercapacitor with the same charge at the same voltage. The potential difference of the battery during a discharge or recharge half-cycle is shown in figure 4.14 [19]. The capacitor voltage decreases linearly, since $C = Q/V$; thus, $dV/dQ = 1/C$. In the case of a battery, the voltages on discharge and recharge are shown as parallel lines of zero slope as a function of charge. The possible difference in two lines during discharge or recharge half-cycles could be due to the polarization of the anode and cathode (ohmic IR

Table 4.1. Types of capacitors and their modes of energy storage. Adopted from [10].

Type	Basic of charge or energy storage	Examples	Battery type
Vacuum	Electrostatic	—	**Primary**
Dielectric	Electrostatic	Mica, Mylar, paper	Leclanché, zinc-MnO_2; alkaline, zinc-MnO_2; Mg-AgCl; Mg-$PbCl_2$; Li-$SOCl_2$; Li CF_x; Al-air; Zn-air
Oxide electrolytic	Electrostatic	Ta_2O_5, Al_2O_3	**Secondary**
Double-layer colloidal electrolyte	Charge separation at electrode interface	Carbon powder and fibers	Lead-acid, Pb-PbO_2; nickel cadmium, Ni-O-OH-Cd; nickel-hydrogen, Ni-O-OH-metal hydride; nickel-zinc, Ni-O-OH-Zn; mercuric oxide-zinc, HgO-Zn; silver oxide(s)-zinc, AgO-Zn; zinc-air; Li-TiS2; Li-MoS_2; Li-CoO_2; Li-C-CoO_2; Na-S; Li-iron sulfides
Redox oxide film	Faradic charge transfer (pseudocapacitance)	RuO_2, MnO_2	
Redox polymer film	Faradic charge transfer (pseudocapacitance)	Polyaniline	
Soluble redox system	Faradic charge transfer (pseudocapacitance)	$Fe(CN)_6^{4-}$, $Fe(CN)_6^{3-}$	

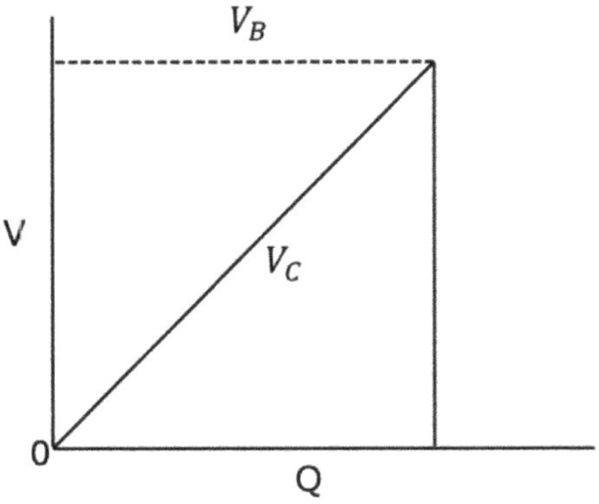

Figure 4.14. Schematic illustration depicting the energies stored by a battery at a constant potential and by a capacitor at a varying potential.

Table 4.2. Characteristics of supercapacitors.

Advantages and disadvantages of energy storage in supercapacitors
Advantages
Long cycle life of > 10^5 to 10^6 cycles
High power density
Can be combined with a battery for hybrid applications
Disadvantages
Limited energy density due to nonfaradic charge storage
Low working potential (3–5 V)

drop or solution resistance). In capacitors, the two lines can also be separated, denoting a significant IR drop. Figure 4.14 shows the voltage vs. charge relationships for a supercapacitor and an ideal battery at constant voltage. The integral VQ under the two lines gives the energy of charging, which for the supercapacitor is half that of the battery. The CV response of a supercapacitor for a single-direction voltage sweep is the mirror image of that generated for the opposite sweep direction if there is no contribution from the diffusion process. In contrast, processes in batteries are rarely reversible, because different ranges of potentials are required for oxidation and reduction [10].

In diffusion-controlled reversible processes, if the reactions are diffusion controlled, the CV does not show a mirror image, because the potential for the anodic peak current may be more positive than that of the cathodic current. Therefore, different potentials are required on the anodic sweep compared to the cathodic sweep to maintain the same diffusion-controlled current produced by the concentration gradient in the diffusion layer. Table 4.2 lists the advantages and disadvantages of supercapacitors.

4.12 Supercapacitor cell balancing

Cell-balancing circuits are used because individual supercapacitors have a leakage current, and their tolerance also varies, along with the probability that the voltage rating may exceed the cell maximum. Thus, a balancing circuit is used to maintain the same level of voltage between all cells, preventing differential aging between cells and protecting them from overvoltage. Supercapacitors in series can be balanced by such circuits more effectively; however, they do not provide an effective solution for applications with high cycling rates, as their currents are higher than the balancing current.

4.13 Supercapacitor manufacture and construction

The supercapacitors used in final applications comprise many combined components, including current collectors, electrodes, an electrolyte, binders, separators, connectors, and a container. The assembly of separators and electrodes can be arranged in a series or parallel configuration within a container to attain higher voltage or higher current.

A binder is required to fix the selected electrode material, such as AC, graphene, carbon nanotubes, etc. to the current collector, such as aluminum foil. The electrodes and current collectors are either stacked or wound with the terminals used for external connections.

Supercapacitor electrodes are prepared by mixing AC with a binder (5%–10% by weight). Commonly used binders include polyvinylidene fluoride and polytetrafluoroethylene. To enhance the electrical conductivity, carbon black is added to the mixture. A slurry of the mixture is formed after dispersion in a solvent (e.g. N-methylpyrrolidone or isopropyl alcohol) following ball milling. The slurry material can be coated on an electrically conductive surface to form a current collector. This process is known as calendaring. The electrode is then dried at high temperature by evaporating the solvent, leaving the binder. The drawback of this process is that the surface area of the activated electrode is reduced by 50%, down to 500–1000 $m^2\ g^{-1}$.

Supercapacitors are often produced commercially by stacking components in a cylindrical form or a coin form. In cylindrical types, the electrode layer is coated onto both sides of a separator. Coin-type supercapacitors are manufactured by pressing the components in a compact container to ensure good electrical contact, as shown in figure 4.9. Pouch cells represent an alternative packaging method.

4.14 Series configuration of supercapacitors

Higher working voltages can be reached using cells connected in series, although it is recommended to operate cells at lower voltages. This reduces the impact on system life due to unbalanced cell voltages. Series-connected cells become open circuited if an open circuit occurs in any series-connected cell. Therefore, this is a serious concern associated with connecting many cells in a long string. In a series-connected module, the tolerance range of a supercapacitor is between −10% and +20% of its nominal capacitance, and the acceptable maximum capacitance difference between individual cells is 30%. Connecting a large number of supercapacitors in series induces higher voltage disparity; therefore, overvoltage may lead to failure. Series-connected modules should be designed to keep the cells within the acceptable tolerance range of capacitance. The weakest cell in a chain of series-connected capacitors controls the performance. In series-connected supercapacitors, the ESR and capacitance are directly related to the number of cells ($ESR_{Total} = ESR_{Cell} \times$ no. of cells, $C_{Total} = C_{Cell}$/no. of cells).

4.15 Parallel configuration of supercapacitors

Connecting supercapacitors in parallel yields higher capacitance than that of a single unit. Cells connected in this way must be matched in terms of capacitance and ESR [20]. For parallel-connected identical cells, the capacitance increases as a multiple of the number of supercapacitors, as follows: ($ESR_{Total} = ESR_{Cell}$/no. of cells, $C_{Total} = C_{Cell} \times$ no. of cells). Commercial supercapacitor banks use several supercapacitors connected in series by an array of parallel links to achieve the required voltage and capacitance.

4.16 Methods of balancing

Passive balancing is used to equalize the voltages of all cells using electronic circuits, especially in applications that can sustain higher leakage (e.g. backup power systems). In this approach, a resistor with a resistance ten times higher than the cell's average leakage current is attached in parallel to each cell.

References and further reading

[1] Kurra N and Jiang Q 2022 Supercapacitors *Storing Energy* (Amsterdam: Elsevier) 383–417

[2] Sharma P and Bhatti T S 2010 A review on electrochemical double-layer capacitors *Energy Convers. Manag.* **51** 2901–12

[3] Ivanov V D 2024 *The Helmholtz Model* (Heidelberg: Springer Science and Business Media)

[4] Sun Z *et al* 2018 Effect of the electronegativity on the electrosorption selectivity of anions during capacitive deionization *Chemosphere* **195** 282–90

[5] Wu J 2022 Understanding the electric double-layer structure, capacitance, and charging dynamics *Am. Chem. Soc.* **122** 10821–59

[6] Vlad A, Singh N, Rolland J, Melinte S, Ajayan P M and Gohy J-F 2014 Hybrid supercapacitor-battery materials for fast electrochemical charge storage *Sci. Rep.* **4** 4315

[7] Xia H, Wang Y, Lin J and Lu L 2012 Hydrothermal synthesis of MnO2/CNT nano-composite with a CNT core/porous MnO2 sheath hierarchy architecture for supercapacitors *Nanoscale Res. Lett.* **7** 33

[8] Fleischmann S *et al* 2020 Pseudocapacitance: from fundamental understanding to high power energy storage materials *Am. Chem. Soc.* **120** 6738–82

[9] Li M *et al* 2021 Experimental protocols for studying organic non-aqueous redox flow batteries *ACS Energy Lett.* **6** 3932–43

[10] Conway B E 1999 *Electrochemical Supercapacitors* (New York: Springer)

[11] Costentin C, Porter T R and Savéant J M 2017 How do pseudocapacitors store energy? Theoretical analysis and experimental illustration *ACS Appl. Mater. Interfaces* **9** 8649–58

[12] Mathis T S, Kurra N, Wang X, Pinto D, Simon P and Gogotsi Y 2019 Energy Storage Data Reporting in Perspective—Guidelines for Interpreting the Performance of Electrochemical Energy Storage Systems *Advanced Energy Materials* **9** 1902007

[13] Wang C, Yang D, Zhang W, Qin Y, Qiu X and Li Z 2024 Engineering of edge nitrogen dopant in carbon nanosheet framework for fast and stable potassium-ion storage *Carbon Research* **3** 20

[14] Wardani V, Rohmawati L, Setyarsih W, Alfarisi D and Subhan A 2020 Analysis of charging/discharging supercapacitor active carbon/rGO based on natural materials *J. Phys.: Conf. Ser.* **1491** 012044

[15] Liu J *et al* 2018 Advanced energy storage devices: basic principles, analytical methods, and rational materials design *Adv. Sci.* **5** 1700322

[16] Li K *et al* 2022 Atomic scale modulation strategies and crystal phase transition of flower-like CoAl layered double hydroxides for supercapacitors *CrystEngComm* **24** 2081–8

[17] Lee S C, Kwon O, Thomas S, Park S and Choi G H 2014 Graphical and mathematical analysis of fuel cell/battery passive hybridization with *K* factors *Appl. Energy* **114** 135–45

[18] Pell W G and Conway B E 1996 Quantitative modeling of factors determining Ragone plots for batteries and electrochemical capacitors *J. Power Sources* **63** 255–66

[19] Sharma P and Kumar V 2021 Investigation of the behaviour of supercapacitors using theoretical models *Physica B: Condens. Matter* **619** 413212

[20] Satpathy S, Das S and Bhattacharyya B K 2020 How and where to use super-capacitors effectively, an integration of review of past and new characterization works on super-capacitors *J. Energy Storage* **27** 101044

[21] Elgrishi N, Rountree K J, McCarthy B D, Rountree E S, Eisenhart T T and Dempsey J L 2018 A practical beginner's guide to cyclic voltammetry *J. Chem. Educ.* **95** 197–206

Chapter 5

Materials for supercapacitors

5.1 Metal oxides

Many transition-metal oxides have been investigated due to their pseudocapacitive supercapacitance in aqueous and nonaqueous electrolytes. The surface-bound water and the structure are the main differences between aqueous and nonaqueous metal oxides [1]. The presence of water on the surface may result in ion adsorption and ion diffusion into the inner surface, as observed for $RuO_2 \cdot nH_2O$. The aqueous electrolytes include MnO_2 (which exists in a number of polymorphs), $RuO_2 \cdot nH_2O$, Mn_3O_3 (spinel structure), and transition-metal hydroxides [2].

5.1.1 Manganese oxide (MnO_2)

The pseudocapacitive compound MnO_2 was first studied by Lee and Goodenough in 1999 in an amorphous form, $MnO_2 \cdot nH_2O$, with an aqueous electrolyte, KCl. In this scenario, charge storage occurs through the redox reactions of Mn between two surface oxidation states, +3 and +4. This results in a rectangular cyclic voltammogram with a capacitance of 200 F g^{-1} according to the reaction: $MnO_2 + xA^+ + xe^- \leftrightarrow A_xMnO_2$, where A is an alkali metal cation. There are also reports of other transitions, such as transitions between different oxidation states, Mn(III)/Mn(II), Mn(IV)/Mn(III), and Mn(VI)/Mn(IV). Compared to RuO_2, MnO_2 is more abundant and cheaper. Furthermore, it has been demonstrated that an ultrathin film of MnO_2 can achieve a specific capacitance of more than 1000 F g^{-1} if all the storage sites in MnO_2 are accessed. The difference between the electrochemical storage behavior of amorphous and crystalline MnO_2 is that charge storage only takes place on the surface in the case of amorphous MnO_2, while bulk ion intercalation occurs in crystalline MnO_2. Due to the greater contribution of faradic reactions, Brousse *et al* reported a higher capacitance of 250 F g^{-1} in crystalline MnO_2 compared to 160 F g^{-1} in amorphous MnO_2. Figure 5.1 shows the capacitance of MnO_2 allotropes as a function of cycle number. The highest capacitance of 297 F g^{-1} is observed in the spinel form of MnO_2. The main challenge that remains in working with MnO_2 is to

doi:10.1088/978-0-7503-5944-3ch5

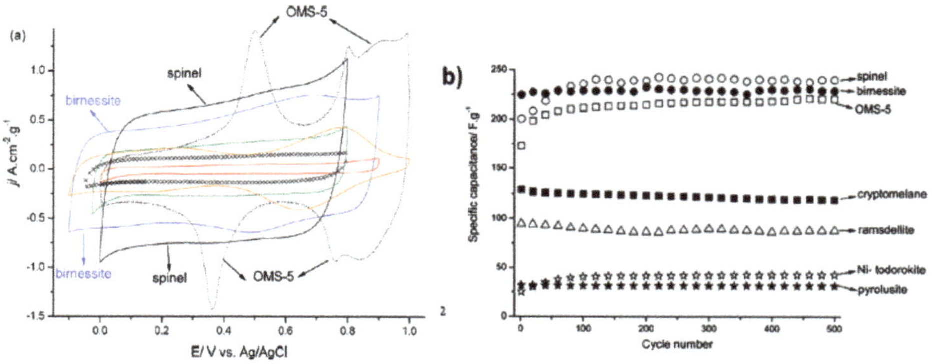

Figure 5.1. (Left) Cyclic voltammogram of a thick composite electrode of amorphous $MnO_2 \cdot nH_2O$ in a KCl electrolyte. Adapted with permission from [4]. Copyright (2020) American Chemical Society. (Right) The specific capacitance and cyclability of different MnO_2 allotropes. Adapted with permission from [4]. Copyright (2020) American Chemical Society.

improve its electrical conductivity in thick films to harness its theoretical capacitance at high charge/discharge rates. Physical and chemical factors such as the microstructure and chemically hydrated state of MnO_2 affect its cyclic stability and specific capacitance. Although highly crystalline MnO_2 offers increased conductivity, it limits proton exchange due to its reduced surface area. The specific capacitances for α-MnO_2 and γ-MnO_2 [3] are between 240 and 320 F g^{-1} under varying electrolytes and current rates, as shown in figure 5.1. Several MnO_2 thin films have been reported to exhibit a specific capacitance of 600 F g^{-1} in a potential window of 0.9–1.2 V in aqueous electrolytes [1].

5.1.2 Ruthenium oxide (RuO_2)

RuO_2, with three oxidation states, was the material historically used to demonstrate pseudocapacitance for the first time. It has a wide potential window and highly reversible redox reactions. Moreover, it provides high thermal stability, a long cycle life, and high electrical conductivity and rate capability. However, unlike MnO_2, its high cost and rarity limit its widespread application in supercapacitors. About 10% of its total charge storage is contributed by double-layer capacitance, although the predominant contribution is due to pseudocapacitance, which results from different reactions in acidic and alkaline media. In acidic electrolytes, the oxidation state of ruthenium changes from (II) to (IV). In alkaline electrolytes, ruthenium oxidizes [5] to RuO_4, RuO_4^{-}, and RuO_4^{2-}. Hence, there are efforts to utilize smaller amounts of RuO_2 while achieving its maximum theoretical capacitance. For example, the use of a slurry method leads to a high capacitance of 750 F g^{-1}, which is 50% of the theoretical capacitance of hydrous ruthenium oxide ($RuO_2 \cdot nH_2O$). Since the main energy storage mechanism in RuO_2 is pseudocapacitance, a more electroactive surface can be obtained by depositing a film of RuO_2 on a rough surface or through the synthesis of RuO_2 nanoparticles. Combining RuO_2 nanoparticles with activated carbon leads to a capacitance of 1340 F g^{-1}, which is closer to the theoretical capacitance due to the effective exposure of the surface to the electrolyte. RuO_2 is also used in combination with other low-cost materials such as an $SnO–RuO_2$ composite, which showed a specific capacitance of 710 F g^{-1} in KOH electrolyte.

5.2 Metal hydroxides

In some transition metals with +3 oxidation state, the additional positive charge is compensated by anions intercalated within the interlayer spaces, thus forming a layered double hydroxide. The nanoscale interspacing allows reversible intercalation of the ions. For example, nickel and cobalt transition-metal hydroxides are used as electrodes in rechargeable batteries. The phase transformation that occurs during charge storage in these materials leads to a constant potential during charging and discharging. In supercapacitors, these hydroxides are used as pseudocapacitive electrodes. The reversible redox reaction in $Ni(OH)_2$ (namely, $Ni(OH)_2 + OH^- \leftrightarrow NiOOH + H_2O + e^-$) occurs via the formation of Ni^{2+}/Ni^{3+} in KOH, which gives a theoretical capacity of 1040 C g^{-1} within a potential window of 0.5 V.

The capacitance of nanostructured $Ni(OH)_2$ is reported to be 1750 F g^{-1} over its entire range of potential, while within smaller potential windows, it can be higher than those of RuO_2 or MnO_2. Some materials, such as lithium ions and sodium ions, have shown pseudocapacitive behavior in nonaqueous electrolytes. The pseudocapacitance in these materials arises due to redox reactions induced by ion insertion, which maintains charge neutrality. The absence of water leads to the degradation of electrochemical cells and prevents them from providing an ionic conduction path. The advantage of nonaqueous electrolytes over materials in aqueous electrolytes is that they provide a wider potential window (up to 1–4 V). Examples of such materials include $T–Nb_2O_5$, TiO_2, $H_2Ti_3O_7$, V_2O_5, MoO_3, etc [2]. Some examples of the cyclic voltammetry performance of various materials are provided in figure 5.2.

5.3 Carbon

Carbon, as the most abundant material, offers low-cost development for any application. Carbon is used in various forms for energy storage in supercapacitors. A rectangular cyclic voltammogram is obtained for carbon electrodes in both organic and inorganic electrolytes. Among carbon materials, activated carbon has a high surface area (>3000 m^2 g^{-1}) and this material is produced through the heat treatment of carbon-rich organic precursors under an inert atmosphere, a process known as carbonization. The activation process results in the formation of a porous structure, leading to a high surface area. In the activation process, the controlled partial oxidation of carbon precursor can be achieved by physical or chemical processes. Chemical activation uses a mixture of carbon and chemicals such as alkalis, carbonates, or acids, while physical activation takes place under an oxidizing atmosphere at high temperatures. Table 5.1 lists the Brunauer–Emmett–Teller (BET) surface area values of activated carbon materials derived from different precursors.

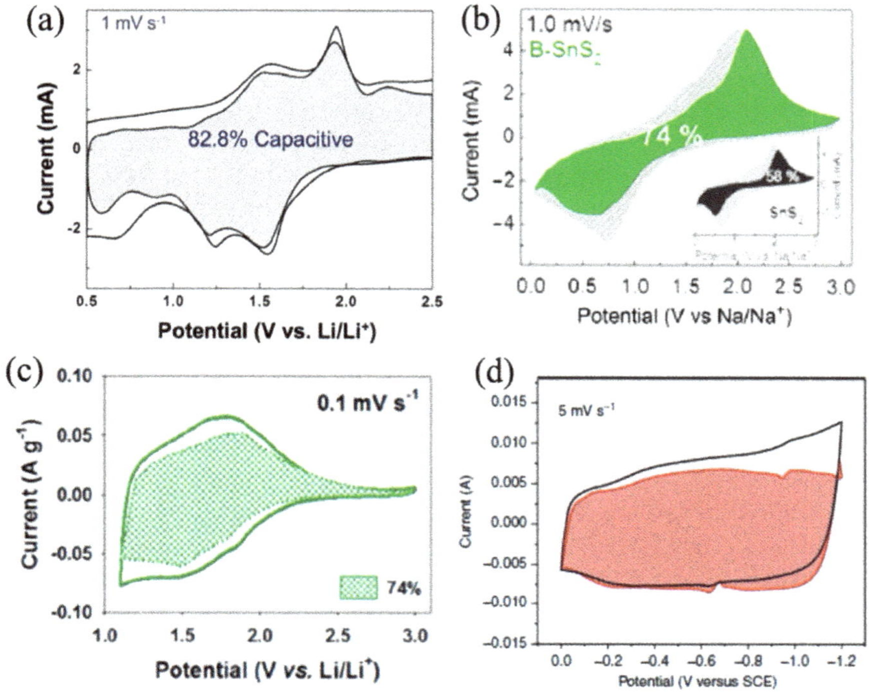

Figure 5.2. (a) Capacitive and diffusion-controlled charge storage contributions in architected MoS_2 cycled in a Li-ion electrolyte at a scan rate of 1 mV s^{-1}. Reproduced from [6]. CC BY 4.0. (b) Cyclic voltammogram showing the capacitive contribution and the diffusion-controlled contribution (gray) of a SnS_2 electrode for Na^+ storage. Reprinted with permission from [7]. Copyright (2016) American Chemical Society. (c) CV curve of T–Nb_2O_5@carbon core–shell material. Reprinted with permission from [8]. Copyright (2015) American Chemical Society. (d) Cyclic voltammogram of FeOOH nanoparticles. Reproduced from [9]. CC BY 4.0.

Table 5.1. BET specific surface areas for different forms of carbon.

Carbon precursor	Activation method	$S_{bet}(m^2\ g^{-1})$
Furfurol	Steam	1040
Coconut shell	KOH	1660
Eucalyptus wood	KOH	2970
Firewood	Steam	1130
Bamboo	KOH	1290
Cellulose	KOH	2460
Potato starch	KOH	2340
Starch	KOH	1510
Sucrose	co_2	2100
Banana fiber	$ZnCl_2$	1100

Nanopores are categorized into different sizes such as micropores (<2 nm), mesopores (2–50 nm), and macropores (>50 nm). Larger pore sizes require longer activation times or higher temperatures.

Another type of carbon is carbide-derived carbon (CDC), which is prepared by the extraction of metals (used as precursors) from carbon at high temperatures. Chlorination and vacuum deposition are common methods used for CDC production. CDC-based supercapacitors show higher capacitance due to their structure. The type of carbide determines the rate performance. Titanium CDC (TiC-CDC) has a maximum gravimetric capacitance of 220 F g^{-1} in KOH and a capacitance of 120 F g^{-1} in organic electrolytes. The SiC-CDC has a maximum volumetric capacitance of 126 F cm^{-3} in KOH and a capacitance of 72 F cm^{-3} in organic electrolytes. Temperatures varying from 600 °C to 1200 °C can be used to tailor the porosity and thus study the effects of pore size. Chmiola *et al* demonstrated that pore sizes of less than 1 nm (smaller than the size of solvated electrolyte ions) lead to higher capacitance.

Single-walled carbon nanotubes (CNTs) and multiwalled CNTs are widely used as electrode materials for microelectronic applications. The specific capacitance of CNTs depends on their purity and their mesoporous character, which is due to the interstitial spaces produced by the entangled nanotubes. The direct growth of CNTs on a conductive substrate reduces the requirement for a binder, thus minimizing the contact resistance between the material and the collector. The specific capacitance of purified CNTs is 20–80 F g^{-1}. Oxidative processes can increase the capacitance to about 130 F g^{-1} by modifying the surface texture and incorporating molecular functionality that produces pseudocapacitance. The chemical activation of CNTs with KOH increased their surface area from 430 to 1035 m^2 g^{-1} and produced both mesoporous and microporous structures. The specific capacitance of the resulting material was 90 F g^{-1} (8.7 μF cm^{-2}) in alkaline media and 65 F g^{-1} (6.2 μF cm^{-2}) in nonaqueous media. As CNTs have limited surface area, they can produce only moderate capacitance. Hence, CNTs are widely used in composites to enhance conductivity.

Another form of carbon, which occurs in planar sheets, is graphene, which has a large surface area (2620 m^2 g^{-1}), flexibility, good electrical and thermal conductivity, and chemical stability. The intrinsic supercapacitance of graphene is reported to be 21 μF cm^{-1}. Graphene-based supercapacitors have a specific capacitance of 75 F g^{-1} in ionic liquids, a capacitance of 135 F g^{-1} in aqueous electrolytes, and a capacitance of 99 F g^{-1} in organic electrolytes. Reduced graphene has a specific capacitance of 205 F g^{-1}. Although graphene provides a high surface-to-volume ratio, a short diffusion distance, the presence of surface functional groups, a wide electrochemical window, and open pores for ion transport, it suffers from irreversible capacity due to the restacking of graphene layers, which reduces its initial coulombic efficiency and surface area and lowers its energy density. In composites, graphene is used with metal oxides due to their complementary synergistic properties. It has been found that the total capacitance of such composites is higher than the sum of the individual contributions. RuO_2 and graphene hybrid materials showed a specific capacitance of 479 F g^{-1}, while the specific capacitance of a graphene–tungsten oxide composite was 143.6 F g^{-1}. The chemical doping of graphene to produce an N-doped graphene-oxide supercapacitor [10] yielded specific capacitances of 242 and 320 F g^{-1} for highly nitrogenated graphene oxide.

Mesoporous carbons are suited for high power output while maintaining significant capacity. Compared to microporosity, which decreases ion mobility and reduces the power handling of electrodes, mesopores provide a wider path for ion transport and thus maintain capacitance at high current densities. Mesoporous carbons are prepared by the carbonization of aerogels or cryogels, in which the carbonization precursors consist of one thermally unstable and one thermosetting component. Other methods include the catalyst-assisted activation of carbon precursors with metal oxides.

5.4 Polymers

Polymers used for supercapacitor applications are electrically conductive (10^4 S cm^{-1}) and highly electroactive. They also change their oxidation state upon the application of an electric field. The typical cyclic voltammogram of a polymer shows a current peak at the redox potential but is otherwise nearly rectangular. The charge–discharge characteristics are battery-like rather than triangular, as shown in figure 5.3 [3].

The common conductive polymers that exhibit pseudocapacitance are polypyrrole (PPy), polyaniline (PANI), and poly(3,4-ethylenedioxythiophene) (PEDOT), which offer ease of preparation and low cost. The charge storage mechanism in conductive polymers involves proton-doping interactions with the material's backbone. However, fast discharge and direct charging in polymer-based supercapacitors lead to eventual swelling and cracking after multiple cycles. A large amount of irreversible reduction occurs due to the relaxation of the polymer matrix, which can

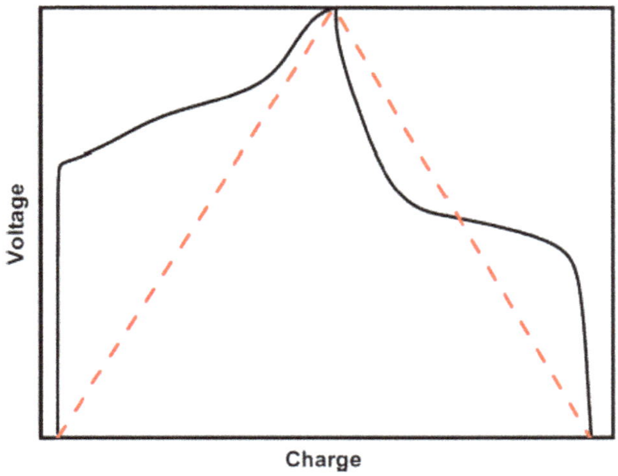

Figure 5.3. Galvanostatic cycling showing the potential limits of an ideal pseudocapacitor cell and a cell based on conductive polymers. Cells using conducting polymers are shown as a black line and the ideal pseudocapacitor as a dashed red line. Reprinted from [11], Copyright (2016), with permission from Elsevier.

be reduced by slower charging. A high bulk capacitance of up to 400 $F g^{-1}$ can be obtained throughout the material, including the surface. The potential window of polymers is limited by the absence of oxidation states and also due to breakdown by overoxidation. In this case, charges are not redistributed fast enough within the polymer matrix. PANI and graphene composites [12] showed a specific capacitance of 1000 $F g^{-1}$ in an aqueous electrolyte. Further improvements in polymer-based supercapacitors can be obtained by controlling nanoscale features with high surface area for increased contact with the electrolyte. This improves charge uptake into the polymer and shortens the path length for faster transport of ions into the polymer backbone, while the presence of more material in a smaller space reduces strain, improving cycle life.

5.5 Electrolytes

Supercapacitor electrolytes should have a high concentration of nearly 0.2 M to avoid the problem of depletion during the charging of the supercapacitor. The ESR of a supercapacitor depends on the temperature coefficient and the conductivity. Moreover, a high potential window, electrochemical stability, low cost, and low toxicity are further requirements for high-performance supercapacitors. Corrosion of the electrode and the current collector also depends on the nature of the electrolyte. The conductivity of aqueous electrolytes is nearly 1 S cm^{-1}. The potential window of aqueous electrolytes limits the cell voltage to 1 V due to water decomposition at 1.23 V. Although organic electrolytes provide a larger potential window of 2.7 V or more, their specific resistance is higher (by almost 20 times) than that of aqueous electrolytes. This results in lower power capability. Aqueous electrolytes provide higher capacitance due to their higher concentration and smaller ionic radius compared to organic electrolytes. Propylene carbonate and acetonitrile are frequently used electrolytes. The organic salts include tetraethylammonium tetrafluoroborate [13], tetraethylphosphonium tetrafluoroborate [14], and triethylmethylammonium tetrafluoroborate [15].

Ionic liquids are pure salts without any solvent, which have a melting temperature of 100 °C. Room-temperature ionic liquids are in a liquid state at ambient temperatures. Such liquids are unusual because they are nonvolatile, noncombustible, and heat resistant. Room-temperature ionic liquids have one organic component and one ionic component to prevent them from forming a stable crystal lattice, and these constituents control the melting point, viscosity, and conductivity. When mixed with polymers, these ionic liquids form ionic liquid gel polymer electrolytes. Ionic liquids can have potential windows of up to 6 V. Figure 5.4 compares a conventional ionic solution to an ionic liquid without a solvation shell.

The main drawback of ionic liquids is their low conductivity of 10 mS cm^{-1}, which is much lower than that of liquid electrolytes. Dilution with an organic electrolyte is used to enhance the conductivity of ionic liquids.

In supercapacitors, the voltage is limited to a specific range by the breakdown of the materials used within the cell. In voltage spectra, the evolution of side redox reactions due to decomposition at high or low voltages causes a sharp current tail, as

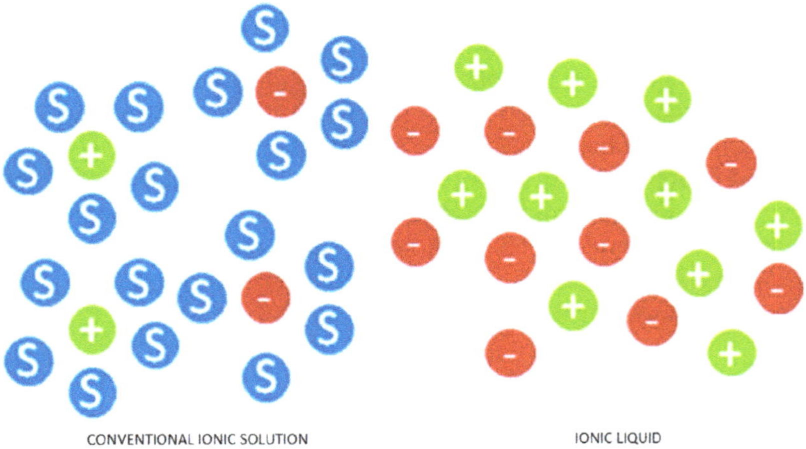

Figure 5.4. Schematic showing the difference between ionic liquids and conventional electrolytes based on dissolved salts. '+' and '−' denote cations and anions, respectively. Reprinted from [11], Copyright (2016), with permission from Elsevier.

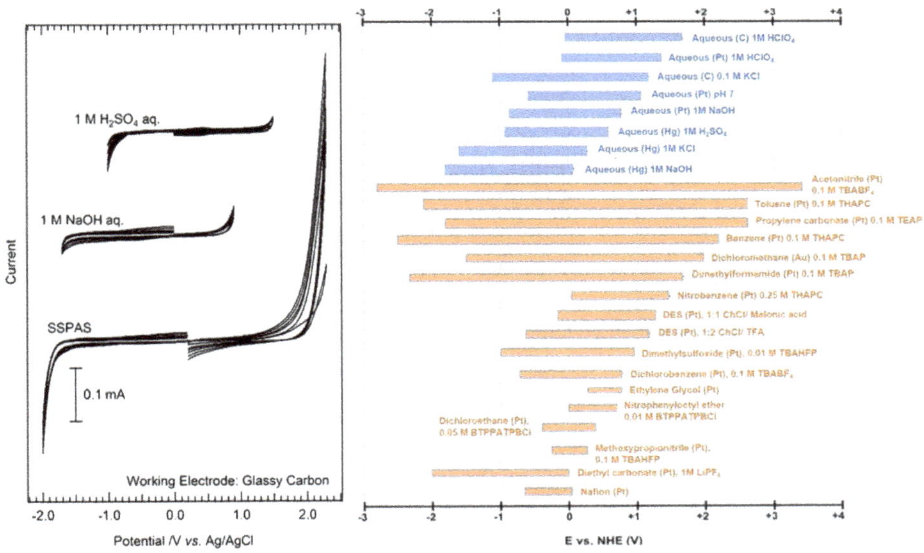

Figure 5.5. (Left) Cyclic voltammogram of a saturated magnesium perchlorate aqueous solution and those of 1 M H_2SO_4 and 1 M NaOH aqueous solutions with a working electrode made from glassy carbon. Reproduced from [16]. CC BY 4.0. (Right) Electrochemical stability windows of representative electrolytes in aqueous, nonaqueous, and deep eutectic solvents. Reprinted with permission from [17]. Copyright (2016) American Chemical Society.

shown in figure 5.5, which can be avoided by controlling the potential window. The redox tails shown in the figure occur when water decomposes due to the presence of a large potential window in a three-electrode cell. A stabilizer can prevent decomposition reactions, increasing the potential window.

Solid-state polymer electrolytes aim to combine gel and solid polymers to function as both the electrolyte and the separator in a compact system without using many components. A microporous polymer matrix (insoluble in the electrolyte) [18] holds the liquid electrolyte via capillary forces, thus creating a solid polymer. Nonpolar polymers such as polytetrafluoroethylene (PTFE), polyvinyl alcohol (PVA), polyvinylidene fluoride (PVDF), and cellulose offer good ionic conductivity [19]. Gel electrolytes can use aqueous, organic, or ionic liquids. Gel polymers offer lower conductance than liquid electrolytes but support efficient ion transport and extended cycle life. Polyvinyl alcohol (PVA) offers good conductance by trapping aqueous electrolytes.

5.6 Separators

The role of the separator is to prevent contact and electron transfer between the anode and cathode and also to provide mechanical stability to the cell. High porosity, wettability, and low tortuosity promote high ionic conductance. The separator must be chemically resistant to corrosion by the electrolyte and any by-products of electrode degradation. The migration of active materials can cause short circuits; therefore, separators must prevent this. Organic electrolyte-based devices use microporous polymers and cellulose paper as separators, while aqueous devices utilize glass, mica, and ceramics. The limitation of paper-based separators is their poor mechanical strength and durability in high-temperature operations. Polyolefin is a microporous separator that serves as an alternative to natural materials due to its high porosity, low cost, flexibility, corrosion resistance, and high mechanical stability. However, its hydrophobic surface limits wetting by the electrolyte; thus, surface treatments are required to make it hydrophilic. There are three main categories of separator structures, namely microporous nonwoven films, gels, and solid polymers [20]. Microporous films have small pores 5–10 nm in diameter and are used for low-temperature applications with organic electrolytes in acidic media.

5.7 Current collectors

The stored electrical charges in supercapacitors are collected and released through current collectors. In most electrode materials, electrical conduction is insufficient, leading to high resistance; thus, efficient contacts and additives are required to facilitate current transport and provide sufficient power. Typical metals or metal alloys used to transfer energy to a common ground in capacitors are aluminum, iron, copper, and steel thin films, which are sometimes produced by pressing materials into a metal mesh. Current conduction occurs between the active material coating and the collector through a low-resistance solder joint. The electron transport distance between active electrode materials and the highly conductive collector should be minimized. Even when there is a good electrical connection at the interface between the collector and the electrode, the active material can exhibit high resistance. The internal resistance of the electrode paste limits its conductivity and prevents the formation of conductive pathways across the thickness of the electrode material. The internal resistance varies depending on the quality of the binding made by the electrode production method. Poor binding significantly affects the operation of the device and may also lead to its failure. Internal resistances are generally reduced by mixing polymeric binding agents such as PTFE, PVDF, Nafion, etc. with conductive carbon additives like acetylene black to increase durability and reduce the degradation of the collector–electrode interface over long periods. These additives increase the internal resistance of the electrode, and for optimum performance, they should contribute 3%–5% of the weight. Dead weight is an important factor in designing electrode materials because non-capacitive weight reduces the performance of the device. Thus, it is suggested to add only a minimal amount of additive paste along with a thin and lightweight current collector. CNTs are an alternative to metal collectors and do not require the use of a binder to achieve sufficient current collection; however, they offer lower energy density compared to AC materials and a higher cost than most metal collectors.

References and further reading

[1] Karandikar P B, Talange D B, Mhaskar U P and Bansal R 2012 Development, modeling and characterization of aqueous metal oxide based supercapacitor *Energy* **40** 131–8

[2] Deka S 2022 Nanostructured mixed transition metal oxide spinels for supercapacitor applications *R. Soc. Chem.* **52** 839–56

[3] Jeong J H, Park J W, Lee D W, Baughman R H and Kim S J 2019 Electrodeposition of α-MnO_2/γ-MnO_2 on carbon nanotube for yarn supercapacitor *Sci. Rep.* **9** 11271

[4] Ghodbane O, Pascal J L and Favier F 2009 Microstructural effects on charge-storage properties in MnO_2-based electrochemical supercapacitors *ACS Appl. Mater. Interfaces* **1** 1130–9

[5] Majumdar D, Maiyalagan T and Jiang Z 2019 Recent progress in ruthenium oxide-based composites for supercapacitor applications *Chem. Electro. Chem.* **6**

[6] Wei X, Lin C-C, Wu C, Qaiser N, Cai Y, Lu A-Y *et al* 2022 Three-dimensional hierarchically porous MoS2 foam as high-rate and stable lithium-ion battery anode *Nat. Commun.* **13** 6006

[7] Chao D, Liang P, Chen Z, Bai L, Shen H, Liu X *et al* 2016 Pseudocapacitive Na-Ion Storage Boosts High Rate and Areal Capacity of Self-Branched 2D Layered Metal Chalcogenide Nanoarrays *ACS Nano* **10** 10211–9

[8] Lim E, Jo C, Kim H, Kim M-H, Mun Y, Chun J *et al* 2015 Facile Synthesis of Nb_2O_5@Carbon Core–Shell Nanocrystals with Controlled Crystalline Structure for High-Power Anodes in Hybrid Supercapacitors *ACS Nano* **9** 7497–505

[9] Owusu K A, Qu L, Li J, Wang Z, Zhao K, Yang C *et al* 2017 Low-crystalline iron oxide hydroxide nanoparticle anode for high-performance supercapacitors *Nat. Commun.* **8** 14264

[10] Li S M *et al* 2015 N-doped structures and surface functional groups of reduced graphene oxide and their effect on the electrochemical performance of supercapacitor with organic electrolyte *J. Power Sources* **278** 218–29

[11] González A, Goikolea E, Barrena J A and Mysyk R 2016 Review on supercapacitors: technologies and materials *Renew. Sustan. Energy Rev.* **58** 1189–206

[12] Zhang K, Zhang L L, Zhao X S and Wu J 2010 Graphene/polyaniline nanofiber composites as supercapacitor electrodes *Chem. Mater.* **22** 1392–401

[13] Dhanda M, Arora R, Ahlawat S, Nehra S P and Lata S 2022 Electrolyte as a panacea to contemporary scientific world of super-capacitive energy: a condense report *J. Energy. Storage.* **52** 104740

[14] Gupta A K, Singh M P, Singh R K and Chandra S 2012 Low density ionogels obtained by rapid gellification of tetraethyl orthosilane assisted by ionic liquids *Dalton Trans.* **41** 6263–71

[15] Ghadimi L S *et al* 2023 A low-concentration 2-step synthesized spiroquaternary ammonium salt electrolyte for commercially stable pouch-type supercapacitors *J. Energy Storage* **69** 107978

[16] Tomiyasu H, Shikata H, Takao K, Asanuma N, Taruta S and Park Y-Y 2017 An aqueous electrolyte of the widest potential window and its superior capability for capacitors *Scientific Reports* **7** 45048

[17] Polcari D, Dauphin-Ducharme P and Mauzeroll J 2016 Scanning Electrochemical Microscopy: A Comprehensive Review of Experimental Parameters from 1989 to 2015 *Chemical Reviews* **116** 13234–78

[18] Yadav N, Mishra K and Hashmi S A 2017 Optimization of porous polymer electrolyte for quasi-solid-state electrical double layer supercapacitor *Electrochim. Acta.* **235** 570–82

[19] Karthika P, Rajalakshmi N and Dhathathreyan K S 2013 Flexible polyester cellulose paper supercapacitor with a gel electrolyte *Chem. Phys. Chem.* **14** 3822–6

[20] Zhang Q *et al* 2022 Swelling-reconstructed chitosan-viscose nonwoven fabric for high-performance quasi-solid-state supercapacitors *J. Colloid Interface Sci.* **617** 489–99

[21] Fleischmann S *et al* 2020 Pseudocapacitance: from fundamental understanding to high power energy storage materials *Am. Chem. Soc.* **120** 6738–82

Chapter 6

Hybrid supercapacitors

As batteries and supercapacitors are becoming the most promising electrochemical energy storage devices, their independent limitations are leading to the development of hybrid devices to achieve the best of both worlds. Hybrid supercapacitors are devices with much improved capacitance and energy storage compared to supercapacitors based on electric double-layer capacitance (EDLC) or pseudocapacitance. This chapter focuses on the fundamental aspects of designing different hybrid supercapacitors based on their classification into different material types.

Supercapacitors are generally classified into three categories based on their storage mechanisms: electric double-layer capacitors, pseudocapacitors, and a combination of these called the hybrid supercapacitor [1]. Ragone plots of the energy/power performances of different energy storage systems show that the performance of supercapacitors lies between those of batteries and conventional capacitors. As mentioned above, supercapacitors provide two-fold advantages of both EDLC and pseudocapacitance due to electrode–electrolyte interface-limited faradic interactions. The high energy density of batteries mainly depends on the intercalation and deintercalation of diffused ions, whereas the energy density in supercapacitors is restricted only by the charge stored on the surface. Hybrid supercapacitors provide higher energy density and capacitance because they combine the properties of EDLC and pseudocapacitance, in a similar way to the combination of batteries and supercapacitors in a circuit. Hybrid materials use a variety of complementary synergistic combinations in a single device to improve overall performance. The components that are considered mainly possess electroactive and conductive natures to amplify the specific capacitance and cell voltage using a combination of faradic and nonfaradic processes. Rectangular cyclic voltammograms ($i = C_{\mathrm{EDL}}\nu$, where $\nu = dV/dt$) and linear gravimetric charge–discharge plots ($V \propto -i_d t / C_{\mathrm{EDL}}$, where i_d is the discharge current) correspond to electrostatic capacitors, i.e. electric double layer based energy storage. The capacitance can be calculated from the slope of the discharge curve at any

doi:10.1088/978-0-7503-5944-3ch6

specific discharge current as follows: $C_{\mathrm{EDL}}(i) = it_d/m\Delta V$, where t_d is the discharge time.

A capacitive cyclic voltammetry response results from the formation of an EDL and multiple redox reactions occurring near the surface of the electrode. The CV curves of pseudocapacitive materials show a symmetrical shape with broad peaks, a small peak-to-peak voltage separation, and a nonlinear discharge curve without a plateau of constant potential. Here, the capacitance can be measured by averaging the slope of the discharge curve using the equation $C(i) = i((dV/dt)_{\mathrm{avg}}{}^{-1})/m$, where $(dV/dt)_{\mathrm{avg}}{}^{-1}$ is the inverse average derivative of the discharge potential window with respect to discharge time. The capacitance can also be evaluated by integrating the CV curves using the following relation: $C = \int IdV/v\Delta V$, where I is the discharge current in the CV response, v is the scan rate, and ΔV is the operating discharge window. Charge storage in pseudocapacitive materials has the following features: (1) potential-dependent charge storage, (2) surface and near-surface charge storage without a limitation on solid-state diffusion, (3) no phase change during CV measurements. In intrinsically pseudocapacitive materials, the short diffusion path length for ions is given by $l = (2Dt)^{1/2}$, where D and t are the diffusion coefficient and reaction time, respectively. Due to their nanostructures, intrinsic pseudocapacitors reduce the diffusion path length for ions and provide charge storage without ion insertion. Battery materials show a contrasting CV response with well-separated voltage peaks due to redox reactions. The characteristic feature of charging and discharging is a plateau of nearly constant potential, which corresponds to the potential at which faradic redox reactions occur [2]. In this case, the faradic current varies as the square root of the potential sweep rate ($i = \nu^{1/2}$).

The target energy density of hybrid supercapacitors is 200–300 Wh kg^{-1}. Their electrodes are mainly composed of carbon materials like graphene, graphite, and activated carbon, along with metal oxides and conducting polymers. Depending on the electrode assembly, hybrid supercapacitors are categorized as symmetric or asymmetric. Assemblies based on two similar supercapacitor electrodes are known as symmetric hybrid supercapacitors, where the electrode materials exhibit both EDLC and pseudocapacitance. Commonly employed asymmetric hybrid supercapacitors have dissimilar electrode assemblies, such as AC and MnO_2 with AC–Ni$(OH)_2$. Improved energy density can be obtained through the use of structures with good electrical conductivity, the maximum number of electrochemically active sites, and high surface area. Hybrid supercapacitors can also be formed using battery-type electrodes; for example, lithium-ion capacitors use electrodes with a lithium-intercalating phase. The proper selection of design and materials can offset problems with individual batteries or supercapacitors. In hybrid systems, battery-type electrodes possess high energy density, while capacitor-type electrodes maintain high power density, as demonstrated by the combination of lithium-intercalated graphite (as a negative electrode) with a porous carbon EDL positive electrode (to form a Li-ion capacitor). The schematic in figure 6.1 shows the charge–potential profiles of a single electrode in a capacitor (a), a single electrode in a battery (b), and an asymmetric (hybrid) capacitor system (c) combining (a) and (b). The profile of a typical

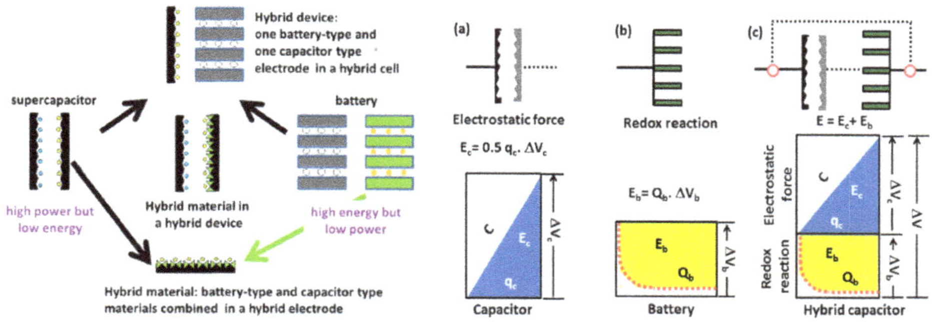

Figure 6.1. (Left) Schematic illustration of different possible hybrids of supercapacitor and battery electrode materials. (Right) Charge–potential profile with corresponding equations for (a) capacitor, (b) battery, and (c) asymmetric capacitor systems. Reproduced from [4] with permission from the Royal Society of Chemistry.

capacitor-like electrode is linear, whereas the redox process in the battery-type electrode shows a phase change, and its potential remains constant during charging and discharging according to the Nernst equation [3] for the phase change. Thus, the energy stored in a capacitor is $\frac{1}{2}(q_c \cdot \Delta V_c)$ (where q_c denotes charge and ΔV_c denotes voltage), which is half of the energy stored in a battery ($=q_b \cdot \Delta V_b$). Hybrid supercapacitors combine the high energy characteristics of a battery with the high power of a capacitor. To increase the stored energy in a hybrid system compared to that of a single electrode, a higher working potential (ΔV) is required at the beginning of the charging process, compared to a capacitor-type electrode, to reach the redox-like potential of the battery-like electrode. Since the battery-type electrode uses a constant electrochemical potential value, the total capacitance of the asymmetric capacitor cannot be reached. Moreover, as it has slower kinetics than an electrostatic capacitor, the battery-type electrode has to perform faster than the capacitor-type electrode [3].

Asymmetric supercapacitors exploit the full potential window of the two different electrodes to achieve the maximum cell operating voltage. The highest cell voltage is obtained when the charges stored in both electrodes are balanced, even under high rate conditions. Since the positive and negative electrodes in hybrid and asymmetric configurations use different charge storage mechanisms, the electrodes exhibit varying rate capabilities as a function of GCD current or CV sweep rate. The dissimilar characteristics of the electrode materials lead to imbalanced charges when a supercapacitor is operated over a wide range of GCD current densities or CV sweep rates. The loss of charge balancing results in the overcharging of one of the electrodes and also in the decomposition of aqueous electrolytes, i.e. water splitting. In the balancing process, the capacitance of electrode materials should first be measured using various mass loadings at different GCD current densities [5]. The stored charge can be calculated based on the loaded mass and GCD current. Then, overlap points can be found where the positive and negative electrodes have the same stored charges at a certain GCD current (not current density). Finally, the

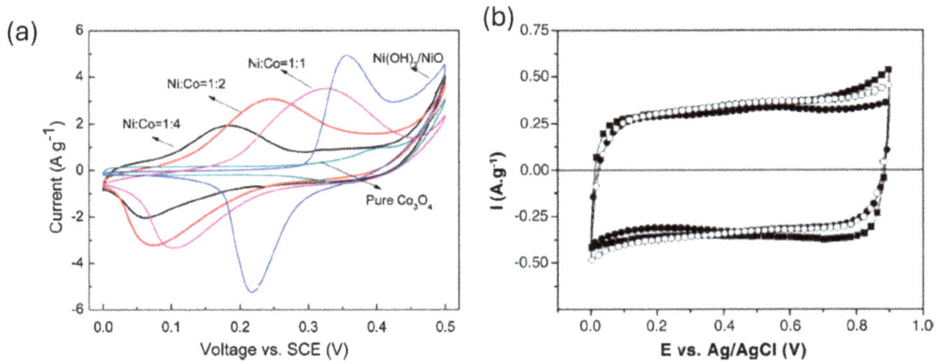

Figure 6.2. Cyclic voltammograms of (a) battery materials and (b) pseudocapacitive materials. Adapted from [6].

selected loading masses and GCD currents can be used to fabricate hybrid and asymmetric supercapacitor devices.

Brousse *et al* described the differences between faradic asymmetric pseudocapacitors and hybrid supercapacitors [5]. Metal oxides such as MnO_2 and RuO_2 display the electrochemical behavior typically observed in capacitive carbon electrodes, which is very different from the behavior observed in nickel oxide or hydroxide and cobalt oxide or hydroxide (figure 6.2). Thus, materials exhibiting electrochemical signatures (e.g. cyclic voltammograms and charge/discharge curves) analogous to that of a battery cannot be classified as pseudocapacitive electrode materials. The metrics for pseudocapacitive materials and faradic behavior are capacitance (F) and capacity (coulombs (C) or mAh), respectively.

Methods for distinguishing pseudocapacitance from capacitance were discussed in an earlier chapter, which quoted the following definition by Conway: 'Regular double layer capacitance arises from the potential-dependence of the surface density of charges stored electrostatically (i.e., non-faradically) at the interfaces of the capacitor electrodes (…). Pseudocapacitance arises at the electrode surfaces where a completely different charge-storage mechanism applies. It is faradic in origin, involving the passage of charge across the double layer, as in battery charging or discharging, but capacitance arises in account of the special relation that can originate for thermodynamic reasons between the extent of charge acceptance (q) and the change of potential (V), so that the derivative of charge acceptance and change in potential, $d(\Delta q)/d(\Delta V)$ or dq/dV, which is equivalent to a capacitance, can be formulated and experimentally measured by dc, ac, or transient techniques (…).' Furthermore, in two-terminal asymmetric or hybrid devices (e.g. those containing a battery-type electrode and a capacitive electrode), the resulting charge–discharge curve of the device may look capacitive because of the triangular profile associated with the capacitive electrode and the plateau associated with the faradic one.

Thus, the pseudocapacitive behavior of the electrode should be investigated using a three-electrode measurement, but a two-terminal device cannot confirm that it is pseudocapacitive. The limited potential window of a carbon electrode in a

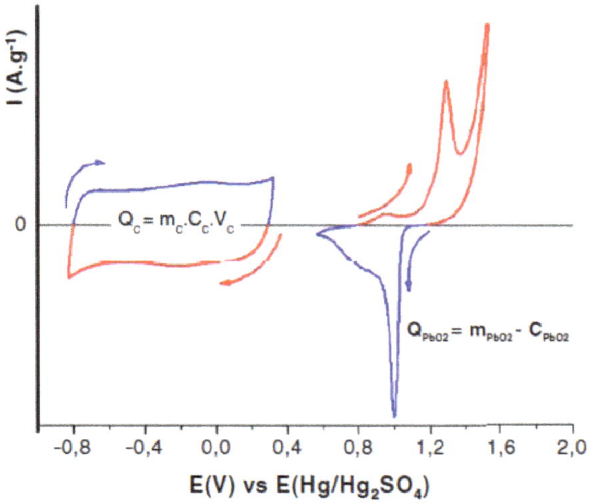

Figure 6.3. The CV response of an AC/PbO_2 hybrid (asymmetric) device in 1 M H_2SO_4. Reproduced from [6].

symmetrical device results in one quarter of the capacitance (F g^{-1}) of a carbon electrode in a three-electrode cell. Thus, by combining an AC electrode (as the cathode) and a battery-type electrode (as the anode), an asymmetric or hybrid device can achieve a complementary potential window, as shown in figure 6.3.

Thus, hybrid devices display higher energy density than those made using symmetric carbon electrodes due to their increased operating voltage and high-capacity faradic electrode. It is clear that the term 'hybrid' supercapacitor is used to describe an assembly consisting of two electrodes with different charge storage behaviors, i.e. one capacitive and one faradic, which results in a device whose operation lies between that of a supercapacitor and that of a battery. 'Asymmetric' supercapacitors either use electrode combinations of the same nature with different masses or two different materials. Brousse *et al* suggested that the term 'asymmetric' should be used only with capacitive or pseudocapacitive electrodes (such as those of AC/MnO_2 asymmetric supercapacitors [7]) to avoid confusion with 'hybrid' devices. Brousse *et al* provided two examples from published data, as shown in figure 6.2. In figure 6.2(a) for $Ni(OH)_2$, a CV scan from 0.0 to 0.3 V versus a saturated calomel electrode (SCE) exhibits a negligible capacity (0 C g^{-1}) or even a capacitance of 0 F g^{-1} due to double-layer capacitance. However, if the same calculations are performed between 0.3 and 0.4 V or 0.4 and 0.5 V versus SCE, the capacity and capacitance vary, reaching values of 300 C g^{-1} and 3000 F g^{-1} or 2000 F g^{-1}. Thus, the capacitance is not constant over the whole potential window, and $Ni(OH)_2$ cannot be considered pseudocapacitive as per reports. Figure 6.2(b) shows the CV response of amorphous MnO_2 [8], which yields a constant capacitance of 150 F g^{-1} over the entire potential window. The cyclic voltammogram of MnO_2 resembles that of the double-layer capacitance produced by AC, but the main origin of this similar signature is faradic reactions involving the electrochemical interconversion of Mn^{4+} and Mn^{3+} and the corresponding insertion of protons from the electrolyte.

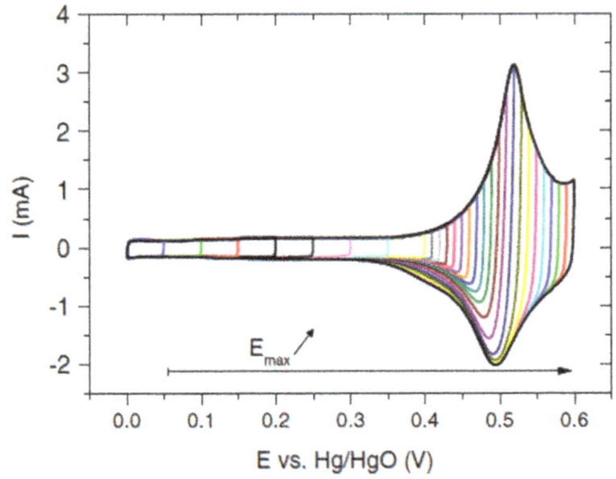

Figure 6.4. Cyclic voltammograms of an electroplated Co_3O_4 thin film annealed at 300 °C in 1 M KOH electrolyte using different upper potential limits. Reproduced from [6].

This process involves the passage of charge across the double layer, which generates a pseudocapacitance. Two-terminal devices may exhibit pseudocapacitor-like behavior from an electrical point of view, but the electrochemical pseudocapacitive response of electrode materials can be measured in three-electrode systems. In another example, electrode behavior appeared to be pseudocapacitive due to a critical size of electrode materials that take the form of a powder or thin film. Due to diffusion over very short timescales, this device showed 'capacitor-like' behavior similar to that of $LiCoO_2$ thin films (6 nm thick) [9]. In this particular case, the capacitor-like behavior was not due to an intrinsic property of the electrode material but was caused by the electrode design or architecture. Dunn *et al* described this as an extrinsic behavior, unlike the behavior of MnO_2, which is intrinsic to the material.

Another example, illustrated in figure 6.4, involves cobalt-based oxides/hydroxides and nickel–cobalt oxides/hydroxides, which exhibit nonconstant capacitance over their potential range. The average capacitance is not available over the whole range of potential; however, the capacity of the electrode (in C g^{-1}) over a desired working range provides a meaningful result for the operation of a full device. The faradic nature of nickel–cobalt oxides or hydroxides [10] may arise due to the high rate capability of asymmetric (hybrid) devices that include double charge layer electrodes such as AC.

Thermodynamically, both asymmetric and hybrid supercapacitors use anodes and cathodes with different potential windows to maximize the working potential window of the cell; thus, the masses of both electrodes need to be balanced to obtain the maximum cell voltage and energy density. Mass balancing consists of finding a suitable mass ratio for the two electrodes. The methodology used to achieve mass balancing is critical not only for asymmetric and hybrid devices but also for symmetric supercapacitors [11]. The reason for this is that the specific capacitance (C_+) and potential window (V_+) of the anode differ from those of the cathode (C_- and V_-). The notion of increasing the specific capacitance by mass balancing is not addressed in this chapter. The following equation is widely used to achieve mass

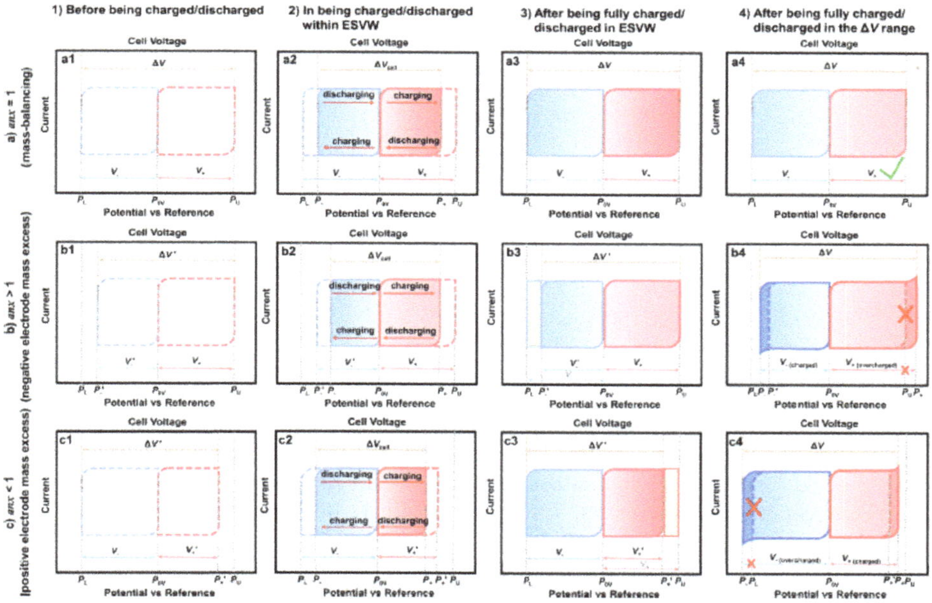

Figure 6.5. Schematic illustration of the CV curves of the positive and negative electrodes of two-electrode supercapacitor devices at different mass ratios of the positive and negative electrodes. Adapted with permission from [12]. Copyright (2020) American Chemical Society.

balancing by balancing the charges at the anode and cathode: $Q_+ = Q_-$ and $C_+ m_+ \Delta V_+ = C_- m_- \Delta V_-$, where the amounts of charge stored in the anode and cathode are Q_+ and Q_-, the gravimetric specific capacitances are C_+ and C_-, m_+ and m_- are the masses, and ΔV_+ and ΔV_- are the potential windows of the anode and cathode (active materials) in the charge/discharge processes, respectively, in a given electrolyte. A complete supercapacitor device is the series combination of a cathode and an anode; thus, its specific capacitance is given by $C_s = \mathrm{C\ m}^{-1} = (C_+ C_-/C_+ + C_-) \times 1/(\mathrm{m}_+ + m_-)$, while its energy density is given by $E = 1/2\mathrm{C}_s(\Delta V)^2$, where $\Delta V = V_+ + V_-$. Thus, the actual potential window ($\Delta V'$) of the device should be evaluated to obtain the correct energy density. For any mass ratio, practically, $\Delta V > \Delta V'$. Figure 6.5 schematically illustrates the effect of the mass ratio on the potential window [11].

Figure 6.6 illustrates the change in the potential of two electrodes during the charging process of an asymmetric cell. A linear increase is observed in the potential of the anode (e.g. carbon) during the charging process, while a faradic (battery-type) interaction takes place at a constant potential at the cathode.

The charge accumulated at the carbon electrode is $Q_+ = m_+ C_+ V_+$, where V_+ is the voltage swing during the charging process, C_+ is the specific capacitance (F g^{-1}), and m_+ is the mass of the electrode. A maximum operational voltage (V_M) defines the maximum stored charge on the surface of carbon. At the cathode, the maximum stored charge is given by $Q_- = m_- c_-$, where m_- and c_- are the mass and specific capacity (Ah g^{-1}), respectively. In an asymmetric supercapacitor, as mentioned

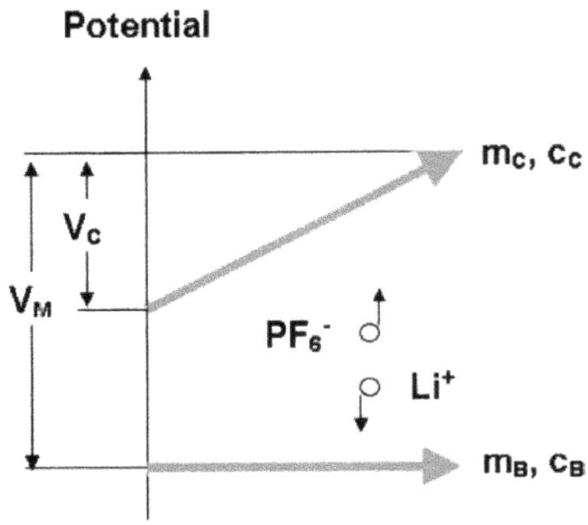

Figure 6.6. Schematic showing the potential changes of two electrodes during the charging process of an asymmetric cell (for example, one with carbon as the positive electrode and a battery-type electrode as the negative electrode). Reproduced from [13].

before, the charges on both electrodes should be balanced, and the ionic charges removed from the electrolyte by the electrodes should also be the same. Therefore, the maximum charge from ions available in the electrolyte is given by $Q_i = (m_i/\rho) c_oF$, where m_i and c_o are the mass and concentration of the electrolyte, respectively; ρ is the density (mass) of the electrolyte, and F is Faraday's constant (=96 484 F mol^{-1}). During the charging process, the change in the charge of the electrodes and the electrolyte are the same; therefore, the maximum charge that can be stored in an asymmetric supercapacitor depends on the smallest value among Q_+, Q_-, and Q_i. If $m_-c_- < m_+c_+V_M$ then the maximum stored charge is controlled by the cathode. The total energy stored is given by

$$E = \int VdC = \left(V_M - \frac{1}{2}V_C\right)m_B c_B.$$

The energy density is the abovementioned total energy divided by the total mass, including both electrodes and the electrolyte. The minimum mass of the electrolyte can be obtained from the mass balance, $Q_i = Q_-$, $m_i = \alpha m_-$, where $\alpha = \rho c_-/c_oF$, which is the minimum value of the mass ratio between the electrolyte and the cathode. Thus, the energy density is given by

$$\varepsilon = \left(V_M - \frac{c_B}{2c_c}\gamma\right)\frac{\gamma c_B}{1 + \gamma(1 + \alpha)},$$

where γ is the mass ratio (m_-/m_+) of the two electrodes. It is clear that the maximum energy density ($\varepsilon_{\max}$) is reached when $\gamma_{\max}$ reaches the value given below at $d\varepsilon/d\gamma/_{\gamma=\gamma_{\max}}=0$:

$$\gamma_{\max} = \frac{1}{1 + \alpha}\left[\sqrt{1 + \frac{2c_c(1 + \alpha)V_M}{c_B}} - 1\right].$$

Furthermore, the swing voltage can be evaluated from the charge balance as follows: $Q_+ = Q_-$, as $V_- = \gamma c_-/c_+$.

It can be concluded that in asymmetric supercapacitors, both the energy density and swing voltage depend on the mass ratio of the electrodes. In another scenario, where the maximum change in an asymmetric supercapacitor is limited by the anode, such that $m_-c_- > m_+c_+V_M$, the swing voltage is V_M and the total stored energy is given by $E = 1/2\ m_+c_+V^2{}_M$. In this case, the minimum mass of the electrolyte can again be obtained from the charge balance, $Q_i = Q_-$ and $m_i = \beta m_+$, where $\beta = \rho c_+/c_oF$, which represents the minimum mass ratio between the electrolyte and the anode (positive electrode). The energy density is given by

$$\varepsilon = \frac{1}{2} c_c V_M^2 \frac{1}{(1+\beta)+\gamma}.$$

Thus, the energy density decreases as the mass ratio of the electrodes increases.

Figure 6.7 shows two examples of asymmetric supercapacitors, AC/$LiPF_6$ EC: DMC/$Li_xTi_5O_{12}$ and AC/$LiPF_6$ EC:DMC/WO_2, to demonstrate the variation of energy density and swing voltage with the mass ratio. Figure 6.8 shows a schematic diagram of various types of supercapacitors discussed earlier. Symmetric supercapacitors [11] generally have electrodes made from EDLC-type materials or materials contributing both EDLC and faradic capacitance [4]. A rare form of

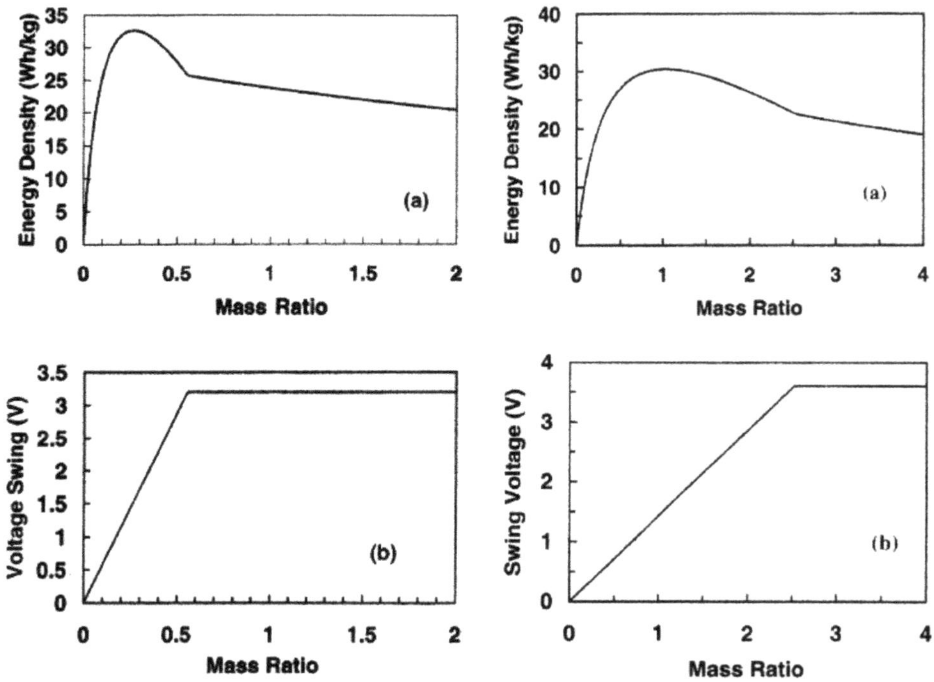

Figure 6.7. Energy density and swing voltage plotted as a function of the mass ratio for the asymmetric supercapacitors AC/$LiPF_6$ EC:DMC/$Li_xTi_5O_{12}$ and AC/$LiPF_6$ EC:DMC/WO_2. Reproduced from [13].

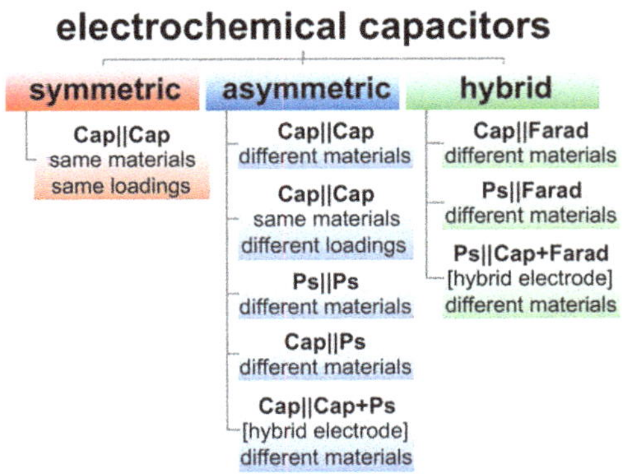

Figure 6.8. Schematic representations of symmetric, asymmetric, and hybrid configurations. The different electrochemical capacitors in the figure can described as: (1) symmetric supercapacitors constructed using the same material for both electrodes e.g. carbon-based materials. Asymmetric supercapacitors may have: (2) two EDLC electrodes of different materials; (3) electrodes of the same material, but with different mass loadings; (4) two pseudocapacitive electrodes with different compositions; (5) one EDLC electrode and a pseudocapacitive electrode, e.g. an AC//MnO_2 aqueous supercapacitor; (6) one EDLC electrode and a composite electrode (e.g. a composite of a carbon-based material with a metal-based species). (7) Hybrid configurations may have a capacitor-type negative electrode and a battery-like positive electrode, and (8) hybrid architectures, e.g. a Li-ion capacitors, may use a battery-type negative electrode and an EDLC positive electrode [16] Copyright © 2024 John Wiley & Sons, Inc.

symmetric supercapacitor is made using only faradic electrodes because faradic reactions can take place either on the cathode or on the anode, but not both. Asymmetric supercapacitors have either different electrode materials (one EDLC and one pseudocapacitive) or the same types of electrode materials with different masses ($m_+ \neq m_-$). It is important to optimize the mass ratio, even for EDLC-based symmetric devices. The reason for this is that, due to the different sizes of the cations and hydrated anions in the electrolyte, the surface areas accessible to the ions are also different. Smaller hydrated ions lead to better kinetics [14]. Thus, merely equal mass loading may not provide the best solution to maximize supercapacitive performance.

Ideally, the potential of zero voltage (PZV) should be in the middle of the capacitive potential range of the device; however, we may note, for example, that figure 6.9 differs from the ideal behavior of AC. Thus, to counterbalance specific capacitance and energy, mass balancing of both electrodes should be performed.

In another example of a symmetric supercapacitor (with the same materials for both electrodes), introducing an appropriate asymmetry into the capacitances of the cathode and anode may widen the operating potential window because the potentials at which the cathode or anode materials start discharging are asymmetric with respect to the PZV, as shown in figure 6.9. Generally, the anode has a narrower potential window than the cathode, which allows the anode to reach the maximum potential before the cathode reaches its maximum limit. Thus, the anode causes the operating voltage to be lower than the sum of the potential windows provided by

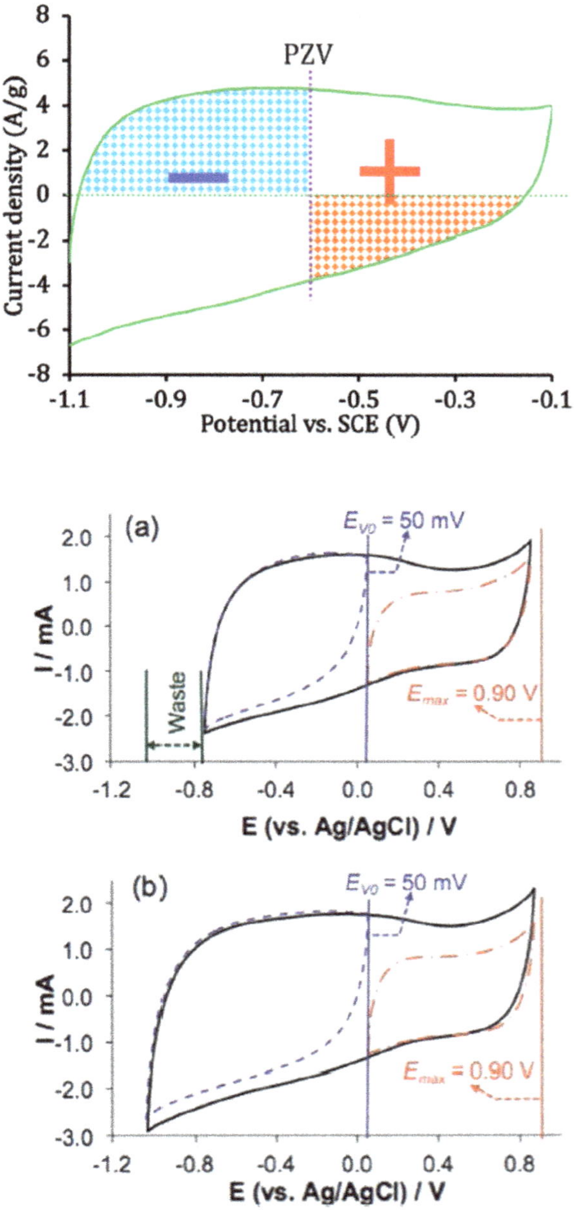

Figure 6.9. (Top) The impact of mass balancing, where the contributions of the positive and negative electrodes to the overall voltammetric signature of the device are indicated with + and − signs, respectively. Reproduced from [15] with permission from the Royal Society of Chemistry. (a) and (b) cyclic voltammograms of a tube cell setup comprising two CMPB-PTFE electrodes and an Ag/AgCl reference electrode. Mass loading: 1.0 mg, electrolyte: 0.3 M K_2SO_4. Electrode capacitance results at 1.6 and 1.9 V are shown for (a) equal mass loading and (b) unequal mass loading (C_+:C_- ratio of 4:3). Reproduced from [15] with permission from the Royal Society of Chemistry.

each electrode. Therefore, it is commonly advisable to measure the open-circuit potential of each electrode, and the capacitance of each electrode should be measured from the open-circuit potential to the limiting potentials of the electrodes. This process allows the electrodes to work across their respective entire operating potential ranges. Again, it is worth mentioning that supercapacitors that use a supercapacitive material and a battery-type material, respectively, for the cathode and anode of a cell, are called hybrid supercapacitors rather than asymmetric supercapacitors. The reason for this is that asymmetric supercapacitors use different materials to gain the advantage of combining two potential windows (varying from 2.0 to 4.0 V) to achieve high energy and high power.

In a two-electrode system, the total capacitance is

$$\frac{1}{C_{\text{tot}}} = \frac{1}{C_-} + \frac{1}{C_+},$$

and the specific capacitances of the individual electrodes are as follows:

$$C_{s_{-/+}} = \frac{C_{-/+}}{m_{-/+}},$$

while the specific capacitance of full cell is

$$C_{\text{S}_{\text{tot}}} = \frac{C_{\text{tot}}}{m_- + m_+}$$

or

$$C\text{S}_{\text{tot}} = \frac{m_+ m_-}{m_+ + m_-} \times \frac{C_+ C_-}{m_+ C_+ + m_- C_-}.$$

Now, in a symmetric device, $m_+ = m_-$ and $C_+ = C_-$; thus, C_s is 25% of the capacitance of each electrode. In an asymmetric cell, $C_+ \neq C_-$, the C_s value of the cell is given by

$$C_{\text{S}_{\text{tot}}} = \frac{m_-}{(m_+ + m_-)} \times C_{s_-}.$$

Because the specific pseudocapacitance is higher than the specific EDLC, $C_- \ll C_+$, the specific capacitance of the asymmetric cell is dominated by the electrode with the lower specific capacitance. In all types of supercapacitors, a lack of effective balancing means that the electrode with the smaller capacity is easily overcharged or overdischarged, while the electrode with the larger capacity is only partially charged or discharged. Thus, unbalanced capacitance between the cathode and anode limits the energy storage performance of the cell. For a cell with equal electrode capacitances, the cell capacitance is given by

$$C_{\text{S}_{\text{tot}}} = \frac{(m_+ + m_-)}{4} \times C_{s_\pm},$$

and the specific energy is given by

$$E_s = \frac{1}{2(m_+ + m_-)} \times C_{S_{tot}} V^2 = \frac{1}{8} C_{S_\pm} V^2.$$

It is recommended that before cell assembly, the individual electrodes should be charged independently, and the cell should then be discharged after assembling the electrodes, which ensures the full utilization of the charge storage capability of each electrode.

Figure 6.10 shows a comparison between the CV and GCD responses of a rechargeable battery, a hybrid supercapacitor, and a supercapacitor in two-electrode and three-electrode configurations. In a battery, the separation between the potentials of the cathode and the anode should be large to enhance the energy of the system. It is evident that the batteries cannot be discharged to a zero potential due to the finite potential difference between the two endpoints of the CV curves. However, the cell voltage becomes zero after the potential on the anode and cathode are the same when a supercapacitor is fully discharged.

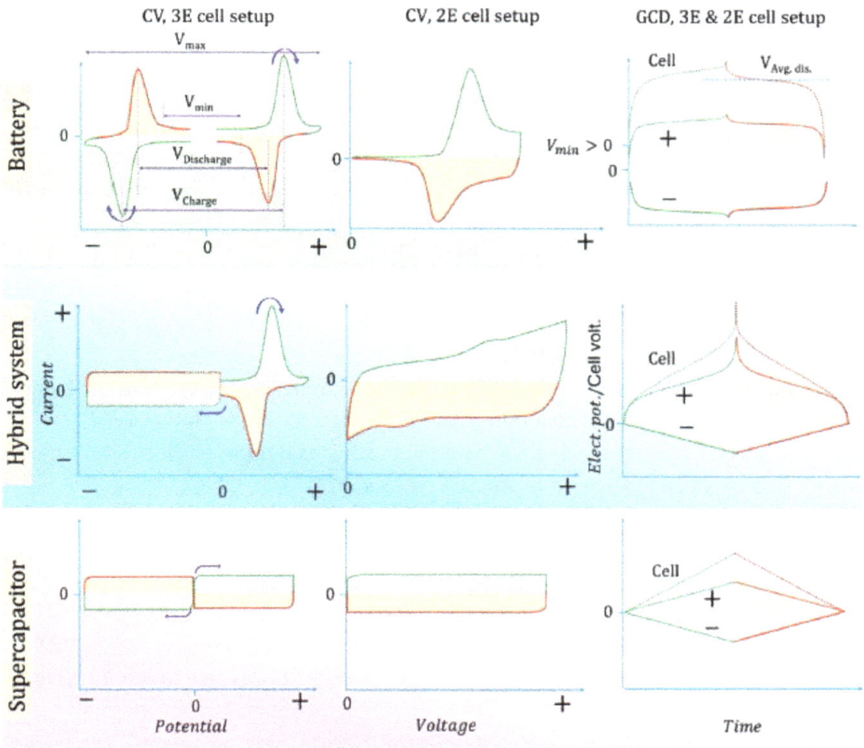

Figure 6.10. Schematic illustrations of the cyclic voltammograms of the positive and negative electrodes in a three-electrode cell setup (left), the cyclic voltammogram of the device for a two-electrode cell setup (middle), and GCD profiles of both two-electrode and three-electrode cell setups (right) of rechargeable batteries (top), hybrid systems (middle), and supercapacitors (bottom). Reproduced from [15] with permission from the Royal Society of Chemistry.

The guidelines for mass balancing are comprehensively outlined by Jing *et al* in their paper [12].

Among the high-performance hybrid supercapacitors are AC//MnO_2, AC//NiO, AC//$Li_2Mn_4O_9$, AC//$LiTi_2(PO_4)_2$, and AC//$Li_4Ti_5O_{12}$ with aqueous electrolytes.

References and further reading

[1] Muzaffar A, Ahamed M B, Deshmukh K and Thirumalai J 2019 A review on recent advances in hybrid supercapacitors: design, fabrication and applications *Renew. Sustain. Energy Rev.* **101** 123–45

[2] Frackowiak E, Meller M, Menzel J, Gastol D and Fic K 2014 Redox-active electrolyte for supercapacitor application *Faraday Discuss.* **172** 179–98

[3] Chowdhury A, Dhar A, Biswas S, Sharma V, Burada P S and Chandra A 2020 Theoretical model for magnetic supercapacitors: from the electrode material to electrolyte ion dependence *J. Phys. Chem.* C **124** 26613–24

[4] Dubal D P, Ayyad O, Ruiz V and Gómez-Romero P 2015 Hybrid energy storage: the merging of battery and supercapacitor chemistries *R. Soc. Chem.*

[5] Li W, Wu M, Yang W, Zhao M and Lu X 2023 Effects of electrode mass loading on the self-discharge of supercapacitors *Electrochim. Acta.* **438** 141550

[6] Brousse T, Bélanger D and Long J W 2015 To be or not to be pseudocapacitive? *J. Electrochem. Soc.* **162** A5185–9

[7] Wang J G, Yang Y, Huang Z H and Kang F 2013 A high-performance asymmetric supercapacitor based on carbon and carbon–MnO_2 nanofiber electrodes *Carbon N. Y.* **61** 190–9

[8] Xu C, Li B, Du H, Kang F and Zeng Y 2008 Supercapacitive studies on amorphous MnO_2 in mild solutions *J. Power Sources* **184** 691–4

[9] Clancy T M and Rohan J F 2018 Ultra-fast cycling of nanoscale thin-film $LiCoO_2$ electrodes in aqueous electrolytes *Chem. Electro. Chem.* **5** 3273–8

[10] Vidhya M S *et al* 2020 Nickel–cobalt hydroxide: a positive electrode for supercapacitor applications *RSC Adv.* **10** 19410–8

[11] Lu C and Chen X 2020 Latest advances in flexible symmetric supercapacitors: from material engineering to wearable applications *Acc. Chem. Res.* **53** 1468–77

[12] Jing L *et al* 2024 The mass-balancing between positive and negative electrodes for optimizing energy density of supercapacitors *Am. Chem. Soc.*

[13] Zheng J P 2003 The limitations of energy density of battery/double-layer capacitor asymmetric cells *J. Electrochem. Soc.* **150** A484

[14] Ohtaki H and Radnai T 1993 Structure and dynamics of hydrated ions https://pubs.acs.org/sharingguidelines

[15] Noori A, El-Kady M F, Rahmanifar M S, Kaner R B and Mousavi M F 2019 Towards establishing standard performance metrics for batteries, supercapacitors and beyond *Chem. Soc. Rev.* **48** 1272–341

[16] Qorbani M, Chen K-H and Chen L-C 2024 Hybrid and Asymmetric Supercapacitors: Achieving Balanced Stored Charge across Electrode Materials *Small* **20** 2400558

Chapter 7

Microsupercapacitors

As modern science and technology continue to advance rapidly, the use of electronic products is also growing at an accelerated pace. To enhance quality of human life, industry is progressively introducing advanced, adaptable, and compact electronic devices and components, including foldable smartphones, health monitors, microsensors, intelligent robots, and other miniature systems [1]. As miniaturized microelectronic devices and autonomous intelligent systems are increasingly used across various applications, there is an urgent demand to develop small energy storage devices that are compatible with them [2–6].

A common challenge in the biomedical and sensor sectors is the frequent need to replace microbatteries, which are limited by their relatively short lifespan. Microbatteries are not suitable for applications requiring high power. While it has been demonstrated that high power can be delivered to integrated systems by connecting batteries in series and/or parallel, this approach leads to an increase in the overall device size. This contradicts the original purpose behind the design of the microsystem [1]. Compared to microbatteries, microsupercapacitors (MSCs) can achieve significantly higher power density within a compact volume without the need for extra integration, making them ideal for flexible systems that demand high power density [1, 7–10].

At present, MSCs are categorized into two structural types: traditional sandwich configurations and in-plane interdigitated structures [11]. Typically, the interdigitated coplanar electrode design demonstrates superior rate capability and greater power density compared to the sandwich configuration. This can be attributed to its unique structure, which significantly reduces the ion diffusion distance of the electrolyte and offers an increased active surface area [1, 12–14]. Based on this, it is essential to review and analyze the impact of device structure on the performance of MSCs.

However, low energy density remains a significant challenge that restricts their practical use. Numerous studies have demonstrated that optimizing the material

doi:10.1088/978-0-7503-5944-3ch7

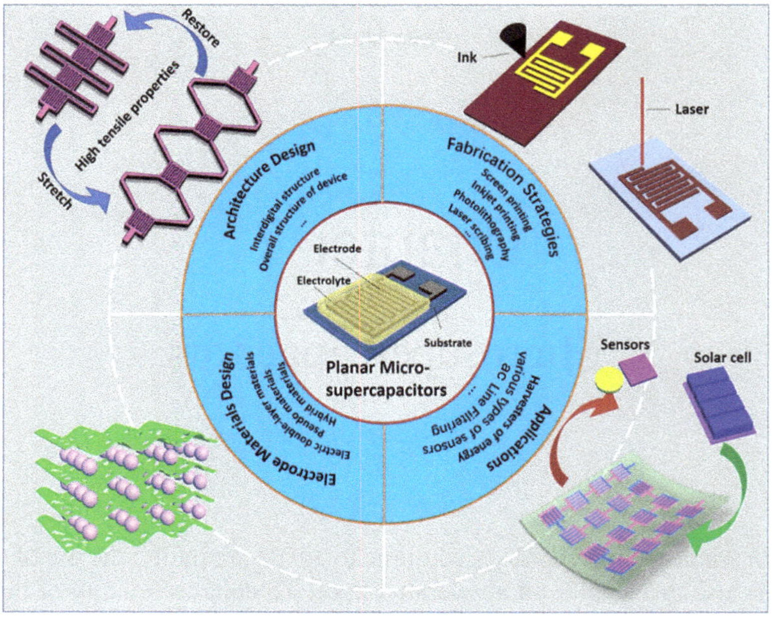

Figure 7.1. Composition of MSCs. Reproduced from [1]. CC BY 4.0.

design can enhance the electrochemical performance of the device [1, 15–17]. Selecting appropriate electrode materials is essential for improving the performance of MSCs, with particular emphasis on the inherent properties of these materials (figure 7.1). To attain excellent electrochemical properties, it is essential to obtain good conductivity, a high surface area for activity, and favorable ion diffusion pathways. To enhance MSC performance, it is essential to optimize the design of all components of the device, including the electrode, electrolyte, substrate, and collector [1].

In this chapter, the latest advancements in MSCs are comprehensively reviewed, focusing on: (i) cutting-edge microfabrication techniques for creating ultracompact and miniaturized MSCs with high resolution, adjustable output voltage, and enhanced capacitance; (ii) precise electrolyte deposition onto predesigned micro-electrode patterns without disrupting nearby devices or electronic components within microelectronic systems; and (iii) practical approaches for developing 3D microelectrodes, and unique electrode structures to improve performance. Additionally, the challenges and potential future directions of research are explored to support ongoing innovation in ultrasmall MSCs for powering microscale electronics.

7.1 Device architecture design

Sandwich-structure MSCs and interdigital-structure MSCs are shown in figures 7.2(a) and (b) [18]. Sandwich MSCs feature a vertical structure with two electrodes and an

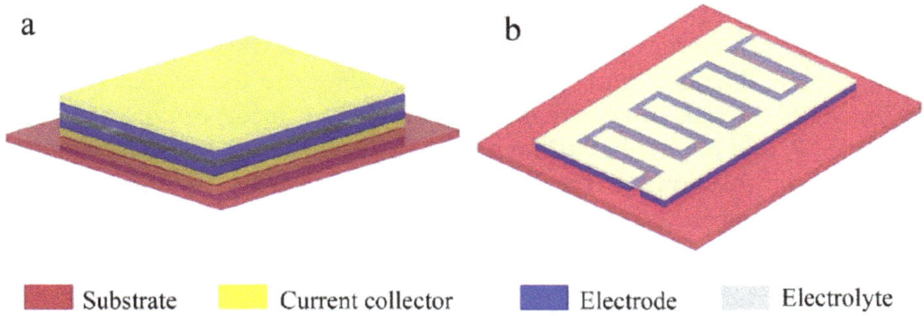

Figure 7.2. Schematics showing MSC configurations. (a) Sandwich type. (b) In-plane type. Adapted from [18], Copyright (2020) with permission from Elsevier.

electrolyte in between, and their performance depends on the thicknesses of the separator and electrodes. Increasing the thickness can enhance active material loading but reduces power density due to longer ion diffusion paths [1, 19]. In contrast, interdigital MSCs have in-plane electrodes separated by an insulating gap, eliminating the need for separators. This design improves ion diffusion efficiency, allows thicker electrodes without compromising power density, and supports better integration in industrial applications [19, 20]. As a result, interdigital MSCs are widely researched for their superior potential in meeting modern demands.

The performance of interdigitated MSCs depends on electrode dimensions, including width, thickness, and gap size. Studies show that increasing the finger width improves current density but may reduce areal capacitance beyond a certain threshold due to longer diffusion paths. Electrode thickness positively impacts capacitance, with optimal designs achieving areal capacitances of up to 0.7 mF cm^{-2} [1]. An all-solid-state MSC, designed with optimized electrode and gap widths, demonstrated an impressive areal capacitance of 7.7 mF cm^{-2} and a volumetric capacitance of 77.3 mF cm^{-3} [21].

Structural optimization, such as adjusting electrode and gap dimensions, can enhance both areal and volumetric capacitances, as demonstrated in recent studies. Narrower gaps improve ion diffusion, while balanced electrode widths avoid excessive diffusion lengths or the loss of active area, thus optimizing power and energy densities. To enhance flexibility for wearable devices and intelligent systems, MSCs with improved tensile properties have been developed [1, 19].

7.2 Device charge storage mechanism

MSCs can be categorized as either symmetric or asymmetric, depending on their energy storage mechanisms. Symmetric MSCs typically utilize electrode materials with a high surface area and excellent conductivity, such as graphene and activated carbon (AC), to store energy efficiently through electrical double-layer capacitance (EDLC). This process involves the rapid and reversible adsorption and desorption of electrolyte ions on the activated surface. Materials composed of carbon, characterized by a high density of micropores and macropores, are highly sought after for use in symmetric MSCs. The pores act as reservoirs for electrolyte ions, facilitating their diffusion and significantly reducing the ion transport distance, which enhances the specific capacitance [18]. Moreover, incorporating heteroatoms such as B, F, P, and N into the carbon matrix can significantly boost capacitance by inducing faradic reactions, altering the Fermi-level density to enhance quantum capacitance, and increasing the interaction with electrolyte ions [18, 22–25]. For example, MSCs based on F-doped and B-doped graphene exhibited outstanding areal capacitances of 17.4 and 16.5 mF cm^{-2}, respectively, which represent a near ten-fold increase compared to undoped graphene MSCs.

Notably, symmetric MSCs have achieved a maximum power density of 1191 W cm^{-3} [26–29]. Nevertheless, achieving high energy densities remains a challenge. To address this, pseudocapacitive nanomaterials such as MXenes, conductive polymers, and transition-metal oxides have been utilized in the development of symmetric MSCs, enhancing their energy storage performance (figure 7.3) [30]. Recent studies have explored asymmetric MSCs, which offer higher energy densities and broader voltage outputs compared to their symmetric counterparts [9, 18, 31]. Zhang and colleagues demonstrated that asymmetric MSCs based on plasma-reduced, nitrogen-doped

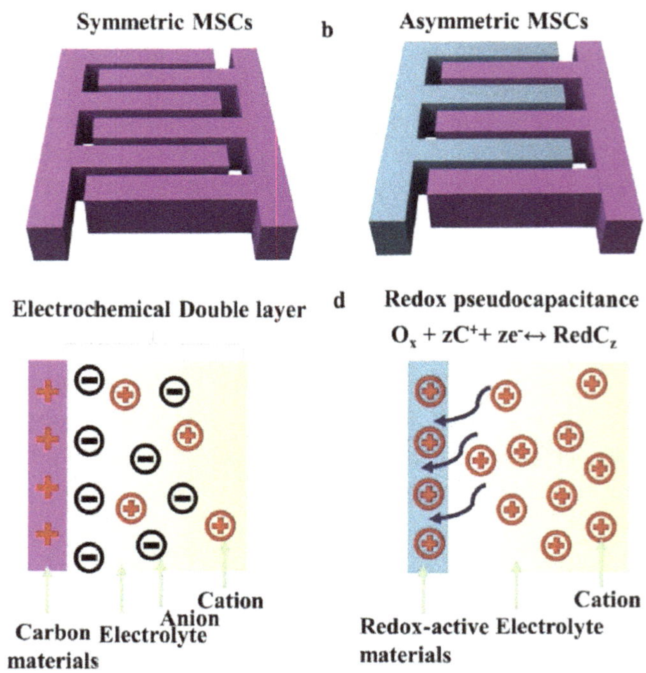

Figure 7.3. Representation of energy storage mechanisms in symmetric and asymmetric MSCs. (a) and (b) Diagrams illustrating the configurations of symmetric and asymmetric MSCs. (c) Electrodes utilizing EDLC mechanisms. (d) Electrodes employing redox pseudocapacitive mechanisms. Adapted from [18], Copyright (2020) with permission from Elsevier.

graphene oxide (PNG) and molybdenum dioxide achieved an impressive energy density of 38.1 mWh cm^{-3} and a broad voltage window of 1.4 V when paired with a polyvinyl alcohol (PVA)/LiCl gel electrolyte. These performance metrics significantly surpass those of graphene-based symmetric MSCs, which typically exhibit energy densities below 10 mWh cm^{-3} and voltage windows of less than 1 V [25]. Moreover, Qin *et al* developed asymmetric VN//MnO_2 MSCs, which exhibited an exceptional energy density of 21.6 mWh cm^{-3} within a 2 V operating window. This performance significantly surpassed the energy densities of approximately 4.9 mWh cm^{-3} observed in symmetric MnO_2 MSCs and 2.2 mWh cm^{-3} in symmetric VN MSCs [32]. A more effective way to enhance the energy density of MSCs involves the development of new types of planar hybrid MSCs, such as zinc-ion microcapacitors (ZIMCs) and lithium-ion microcapacitors (LIMCs) [33–35].

Figure 7.4 illustrates the design and operational principles of high-voltage on-chip MXene-based MSCs. The stability window of Ti_3C_2Tx (an MXene) in a Na_2SO_4 electrolyte is limited to the range of 0.3–0.6 V (figure 7.4(b)). Any increase in the positive potential beyond 0.3 V results in elevated CV curves, indicating excessive polarization. The use of an asymmetric structural design that incorporates an AC positive electrode restricts the $Ti_3C_2T_x$ to functioning solely within the negative potential range of 0.8–0 V. Through the combined effects of various strategies, asymmetric on-chip MSCs utilizing $Ti_3C_2T_x$ and AC are capable of reaching a voltage of 1.6 V, significantly surpassing the 1.2 V achieved by symmetric MXene-based MSCs. This significant improvement in working potential inherently leads to a simultaneous increase in energy and power densities, which is crucial for advancing the miniaturization of microelectronics such as integrated circuits.

To investigate the stable potential window of MXene-based on-chip MSCs, a three-electrode setup was used, with a Pt foil serving as the counter electrode (CE) and an Ag/AgCl electrode as the reference. Figure 7.4(b) illustrates that the cyclic voltammetry (CV) profiles of both the MXene and AC electrodes display nearly

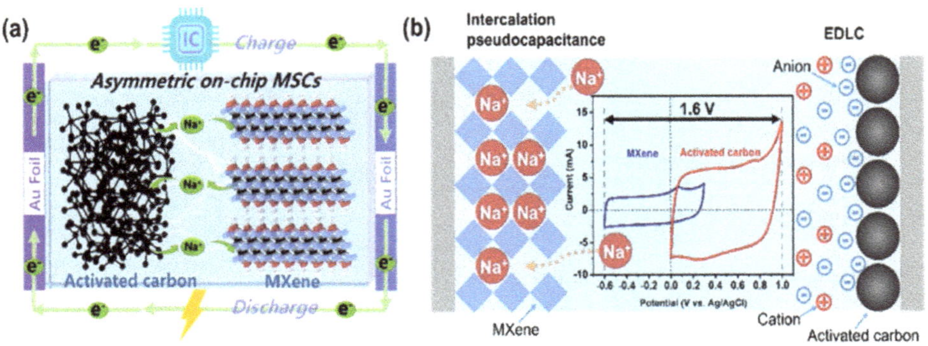

Figure 7.4. Development and fabrication of high-voltage MXene-based on-chip MSCs: (a) Schematic representation of the internal structure and working mechanism of on-chip MSCs. (b) Cyclic voltammetry (CV) profiles of the MXene negative electrode and activated carbon (AC) positive electrode recorded at a scan rate of 10 mV s^{-1} in a three-electrode setup, with the production of high voltage elucidated through an analysis of the charge storage process. Reprinted from [36], Copyright (2020) with permission from Elsevier.

rectangular shapes. However, a small peak is observed at around 0.1 V for the MXene, which is attributed to the intercalation pseudocapacitance that arises when ions are inserted into the layered crystalline structure [36, 37].

7.3 Microsupercapacitors compared to microbatteries

MSCs demonstrate enhanced electrochemical performance compared to that of rechargeable microbatteries (MBs), showing higher power density, improved rate capability, and longer cycle life due to the rapid surface reactions occurring at the electrodes [28, 38–40]. MSCs are often limited by their low energy density, which hinders their broader application. The charge storage mechanism of MBs, on the other hand, involves the insertion and extraction of ions within the electrode material. While this bulk reaction can attain a high areal energy density, it typically leads to a low areal power density, primarily due to the slow ion insertion and extraction kinetics of ions with large ionic radii [41, 42]. A common challenge is that neither conventional MSCs nor MBs can meet the requirement for both high energy and power density at the same time. To leverage the benefits of both energy storage technologies, hybrid micro-supercapacitors (HMSCs), which integrate an ion-intercalation battery electrode with an ion-adsorption capacitor electrode, have been recently introduced [33, 35, 43]. The energy density of these hybrid devices is primarily governed by the number of ions that the capacitor-type electrode can effectively absorb and desorb [44].

Figure 7.5 shows a novel interlayer structural engineering strategy tailored for MXene-based capacitor-type electrodes in HMSCs. This method effectively optimizes ionic and electronic transport within the layered MXene framework, substantially boosting the energy density to match levels typically observed in MBs [44].

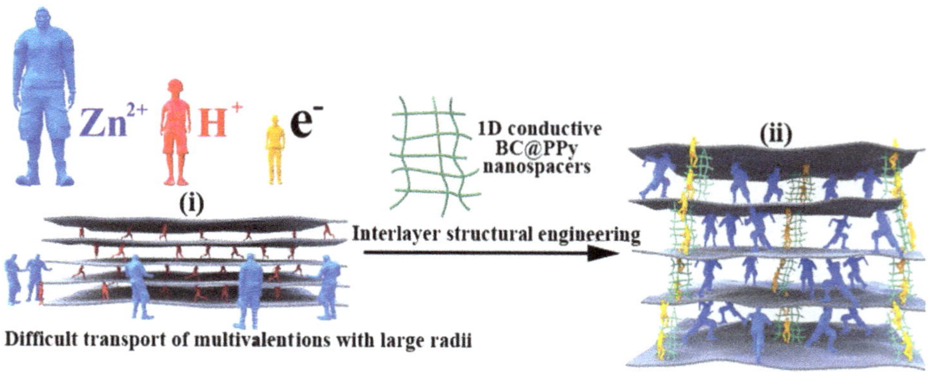

Figure 7.5. The proposed interlayer structural engineering of MXene hybrid films is depicted as follows: (i) the intercalation of multivalent ions with larger ionic radii, such as Zn^{2+}, poses significant challenges compared to the relatively easier insertion of smaller ions like hydrogen. (ii) The introduction of one-dimensional nanospacers expands the spacing between MXene nanosheets, facilitating the swift and reversible intercalation/deintercalation of larger ions. Additionally, these nanospacers improve electron transport across the loosely packed MXene nanosheets. Reproduced from [44]. CC BY 4.0.

7.4 Microsupercapacitor fabrication methods

Various techniques have been established for the fabrication of planar MSCs with interdigitated patterns, as depicted in figure 7.9. These methods allow for the design of the interdigitated structure in planar MSCs, followed by the selection of appropriate electrode materials and electrolytes. This section provides an overview of several widely used fabrication methods, including photolithography [45], ink printing [46], screen printing [47], laser processing [48], plasma etching [49], stamping [50], 3D printing [29], and other approaches [51].

7.5 Electrochemical measurements

A concise overview outlining the progression of microfabrication techniques for creating ultrasmall MSCs is shown in figure 7.6. The two-electrode setup consists of a positive electrode and a negative electrode, both submerged in an electrolyte solution. A voltage is applied between the electrodes using an electrochemical workstation or a power supply. As a result, the recorded voltage represents the total cell voltage. The three-electrode setup consists of a working electrode (WE), a CE, and a reference electrode (RE). Under ideal conditions, current is restricted to flowing between the WE and CE, while the WE's voltage is measured relative to the RE [52].

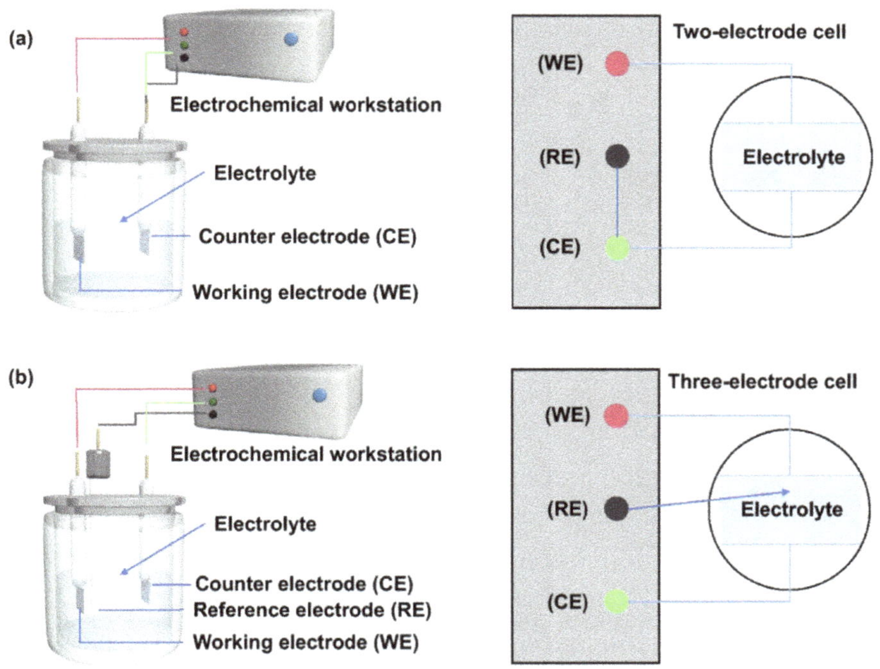

Figure 7.6. Experimental (a) two-electrode and (b) three-electrode setups. Reproduced from [52]. CC BY 4.0.

7.6 Devices in parallel and series configurations

The scalability of $Ti_3C_2T_x$-CNT-based microsupercapacitors is validated through the assembly of multiple devices in series and parallel configurations. When three devices are connected in series, the output voltage range expands from 0.8 to 2.4 V without altering the discharge duration (figures 7.7(A)–(C)). When three devices are connected in parallel, the discharge time increases by a factor of three compared to that of a single device at the same current density of 1 mA cm^{-2} (figures 7.7(D)–(F)), demonstrating the device's excellent scalability for practical applications [53].

7.7 Advantages of 2D and 3D electrodes

While 2D-printed MSCs offer a straightforward and cost-effective fabrication approach, their capacitance is restricted by the difficulty of increasing the active material load without compromising performance. To address this issue, researchers have explored the use of 3D electrodes created through 3D patterning. Studies have demonstrated that graphene-based materials can form three-dimensional structures through the stacking or aggregation of graphene sheets driven by strong π–π interactions. The interface between graphene sheets generates high electrical contact resistance, which is detrimental to the performance of a 3D electrode. Graphene synthesized through chemical vapor deposition has recently demonstrated promise in addressing the issue of contact resistance [54].

Another technique, where constructive interference patterns create localized areas of light intensity that surpass the polymerization threshold, resulting in the formation of a three-dimensional structure. The polymer can be transformed into

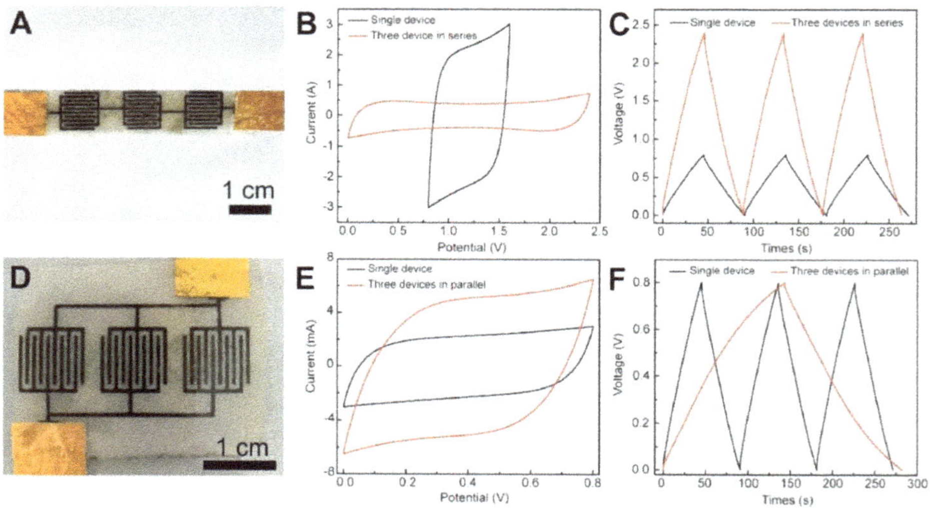

Figure 7.7. Electrochemical performance of a single MSC device is analyzed and compared with series and parallel configurations using the following: (A), (D) digital photographs, (B), (E) CV curves, and (C), (F) constant-current charge–discharge curves. Reprinted from [53], Copyright (2021) with permission from Elsevier.

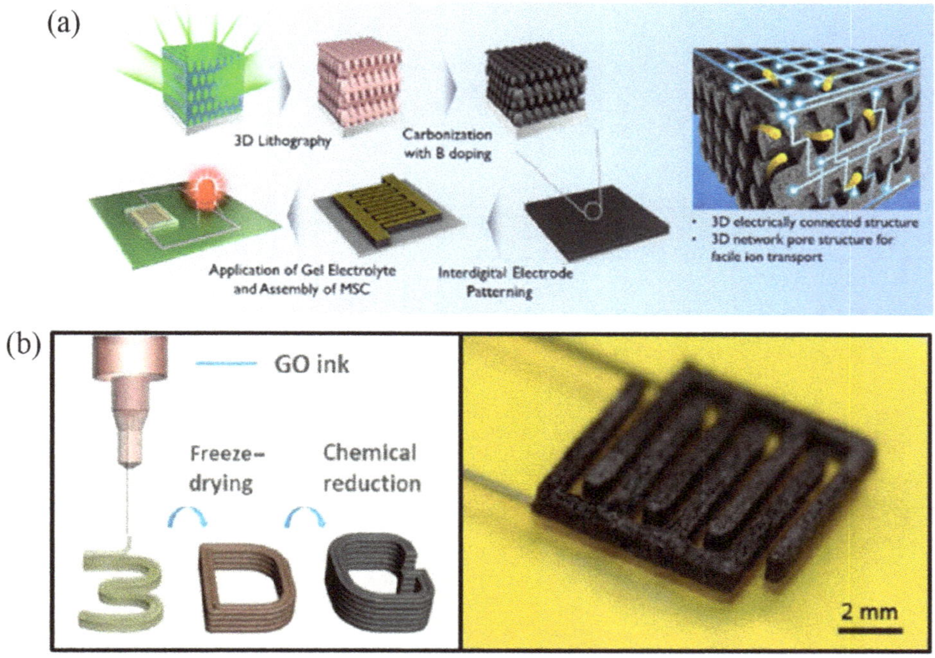

Figure 7.8. (a) MSCs with 3D architected electrodes produced using interference lithography/pyrolysis. Adapted from [31], Copyright (2018) with permission from Elsevier. (b) Printing of reduced graphene oxide (GO) enables the direct fabrication of MSCs with enhanced electrochemical performance. Adapted from [55] with permission from the Royal Society of Chemistry.

carbon-rich material through pyrolysis at approximately 900 °C, enabling its application as electrodes for MSCs. For the procedure illustrated in figure 7.8(a), gold is deposited onto the carbonized electrode serving as the current collector, followed by reactive ion etching to form an interdigitated electrode structure [54]. Extrusion printing (EP), also known as direct ink writing (DIW), is a highly promising 3D printing technique that offers benefits such as versatile material compatibility, affordability, and a straightforward printing process that typically does not require postprocessing. EP enhances the breadth of material choices for MSC electrodes, enabling the utilization of high-surface-area options such as CNTs, graphene, and pseudocapacitive materials [55]. This approach effectively leverages the physical benefits of 3D printing in conjunction with those of EP. Figure 7.8(b) illustrates a study where a graphene oxide (GO) ink prepared with deionized (DI) water was applied to Cr/Au-coated polyimide film to form a current collector. Following the 3D printing process, the resulting structures underwent freeze-drying for 12 h at temperatures ranging from −50 °C to 20 °C [54].

7.8 Electrolytes for microsupercapacitors

The electrolyte is an important element in a supercapacitor, enabling charge balance across its electrodes. The electrolyte used in a supercapacitor cell directly affects its

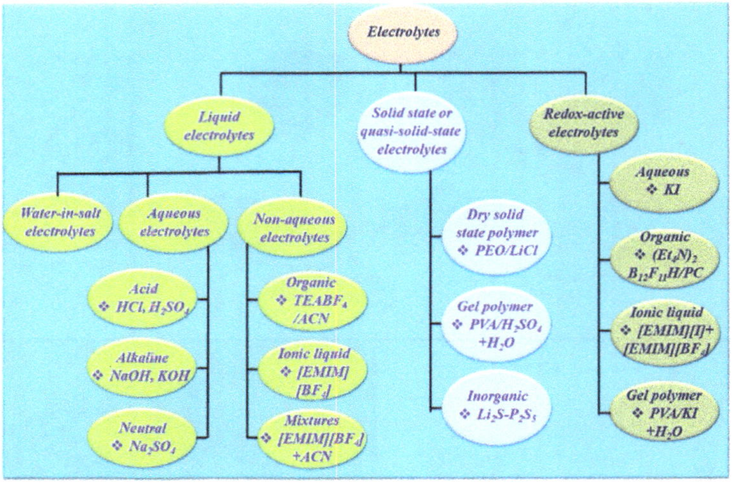

Figure 7.9. Classification of electrolytes for electrochemical supercapacitors. Reproduced from [56] with permission from the Royal Society of Chemistry.

electrochemical performance. An ideal supercapacitor electrolyte should be chemically, electrochemically, and thermally stable; it should be highly conductive, cost-effective, non-flammable, and eco-friendly; it should also have a wide potential range and operate over a broad temperature range [56].

Achieving all the aforementioned properties within a single electrolyte is extremely challenging. Ongoing research continuously aims to improve electrolyte performance, leading to the development and use of numerous electrolytes in existing research literature [57].

The classification of electrolytes for supercapacitors includes liquid electrolytes, solid/quasi-solid-state electrolytes, and redox-active electrolytes, as illustrated in figure 7.9 [56].

Liquid electrolytes consist of specific amounts of salt and liquid. They can be classified into three main types: water-in-salt electrolytes, aqueous electrolytes, and non-aqueous electrolytes [58]. These electrolytes, such as potassium hydroxide (KOH), sodium sulfate (Na_2SO_4), H_2SO_4, and ammonium chloride (NH_4Cl), are widely utilized in electrochemical device development due to their excellent conductivity, mobility, affordability, and easy accessibility. These properties make them highly attractive for energy applications, particularly in supercapacitors. Aqueous electrolytes offer high performance owing to their elevated ionic concentration and efficient ion infiltration into electroactive materials, facilitated by their smaller ionic radius. However, their drawbacks include low voltage stability and a narrow voltage window, significantly limiting energy and power density values [59]. In comparison to KOH, Na_2SO_4 offers a wider potential window in three-electrode setups.

Electrolytes are essential for the development of flexible, wearable, and portable electronics when used in solid or semisolid forms. These electrolytes offer several advantages, such as eliminating the need for separators and addressing the leakage concerns often associated with liquid electrolytes. This simplifies the assembly process of electronic devices [56]. Liquid electrolytes pose challenges for portable electronics due to their low thermal stability and propensity for leakage. Figure 7.9 illustrates three primary types of solid-state electrolytes (SSEs). Among these, gel polymer electrolytes (GPEs) stand out as quasi-solid-state electrolytes due to their single liquid phase. GPEs serve as a viable option to enhance the thermal stability range while providing flexibility and facilitating simplified packaging for electronic devices. Various electrolytes are tetraethylammonium tetrafluoroborate ($TEABF_4$)/acetonitrile (ACN), methylimidazolium tetrafluoroborate ($EMIMBF_4$)/ ACN, polyethylene oxide (PEO) combined with lithium chloride (LiCl) and polyvinyl alcohol (PVA) and potassium iodide (KI) etc.

GPEs are recognized as superior to solid electrolytes due to their enhanced ionic conductivity, while they avoid the safety issues associated with liquid electrolytes. They contribute to the development of versatile supercapacitors, with recent research focusing on enhancing ion conductivity, mechanical characteristics, and overall electrochemical performance. GPEs are widely adopted as solid-state electrolytes in flexible ECs due to their advantages, such as flexibility, high ion conductivity, lightweight nature, and robust mechanical properties [60]. Aziz *et al* extensively explored theories related to various polymer electrolytes, including GPEs and composite polymer electrolytes. In the construction of supercapacitor devices with aqueous electrolytes, challenges such as leakage, stability issues, and voltage limitations are encountered. Gel electrolytes offer enhanced cycling stability and voltage performance compared to aqueous electrolytes.

Supercapacitors' specific capacitance can be enhanced using redox-active electrolytes to introduce pseudocapacitive effects. Pseudocapacitance can arise from both the electrode material, which exhibits pseudocapacitive behavior, and the electrolyte, which contains redox mediators. This not only boosts the specific capacitance but also accelerates faradic reactions at the pseudocapacitive electrode material by providing an electron buffer source.

References and further reading

[1] Bu F, Zhou W, Xu Y, Du Y, Guan C and Huang W 2020 Recent developments of advanced micro-supercapacitors: design, fabrication and applications *Npj Flex. Electron.* **4** 31

[2] Zhao J and Burke A F 2021 Review on supercapacitors: technologies and performance evaluation *J. Energy Chem.* **59** 276–91

[3] Chen D, Lou Z, Jiang K and Shen G 2018 Device configurations and future prospects of flexible/stretchable lithium-ion batteries *Adv. Funct. Mater.* **28** 1–21

[4] Wang Y and Xia Y 2013 Recent progress in supercapacitors: from materials design to system construction *Adv. Mater.* **25** 5336–42

[5] Kyeremateng N A, Brousse T and Pech D 2017 Microsupercapacitors as miniaturized energy-storage components for on-chip electronics *Nat. Nanotechnol.* **12** 7–15

[6] García Núñez C, Manjakkal L and Dahiya R 2019 Energy autonomous electronic skin *Npj Flex. Electron.* **3**

[7] Yue C *et al* 2016 High stability induced by the TiN/Ti interlayer in three-dimensional Si/Ge nanorod arrays as anode in micro lithium ion battery *ACS Appl. Mater. Interfaces* **8** 7806–10
[8] Liu L, Weng Q, Lu X, Sun X, Zhang L and Schmidt O G 2017 Advances on microsized on-chip lithium-ion batteries *Small* **13** 1–12
[9] Ren J, Li L, Chen C, Chen X, Cai Z, Qiu L, Wang Y, Zhu X and Peng H 2013 Twisting carbon nanotube fibers for both wire-shaped micro-supercapacitor and micro-battery *Adv. Mater.* **25** 1155–9
[10] Salian G D, Koo B M, Lefevre C, Cottineau T, Lebouin C, Tesfaye A T, Knauth P, Keller V and Djenizian T 2018 Niobium alloying of self-organized TiO2 nanotubes as an anode for lithium-ion microbatteries *Adv. Mater. Technol.* **3** 1–8
[11] Qi D, Liu Y, Liu Z, Zhang L and Chen X 2017 Design of architectures and materials in In-Plane Micro-supercapacitors: current status and future challenges *Adv. Mater.* **29** 1–19
[12] Yoo J, Byun S, Lee C W, Yoo C Y and Yu J 2018 Precisely geometry controlled microsupercapacitors for ultrahigh areal capacitance, volumetric capacitance, and energy density *Chem. Mater.* **30** 3979–90
[13] Pech D, Brunet M, Mai T, Armstrong K, Gaudet J and Guay D 2013 Influence of the configuration in planar interdigitated electrochemical *J. Power Sources* **230** 230–5
[14] Shi X, Wu Z S, Qin J, Zheng S, Wang S, Zhou F, Sun C and Bao X 2017 Graphene-based linear tandem micro-supercapacitors with metal-free current collectors and high-voltage output *Adv. Mater.* **29** 1–9
[15] Mirvakili S M and Hunter I W 2017 Vertically aligned niobium nanowire arrays for fast-charging micro-supercapacitors *Adv. Mater.* **29** 1–6
[16] Han J *et al* 2015 On-chip micro-pseudocapacitors for ultrahigh energy and power delivery *Adv. Sci.* **2** 1–7
[17] Li X, Yang X, Xue H, Pang H and Xu Q 2020 Metal–organic frameworks as a platform for clean energy applications *Energy Chem.* **2** 100027
[18] Zhang L, Liu D, Wu Z S and Lei W 2020 Micro-supercapacitors powered integrated system for flexible electronics *Energy Storage Mater.* **32** 402–17
[19] Li H and Liang J 2020 Recent development of printed micro-supercapacitors: printable materials, printing technologies, and perspectives *Adv. Mater.* **32** 1–19
[20] Mendoza-Sánchez B and Gogotsi Y 2016 Synthesis of two-dimensional materials for capacitive energy storage *Adv. Mater.* **28** 6104–35
[21] Chih J K, Jamaluddin A, Chen F, Chang J K and Su C Y 2019 High energy density of all-screen-printable solid-state microsupercapacitors integrated by graphene/CNTs as hierarchical electrodes *J. Mater. Chem.* A **7** 12779–89
[22] Hao J, Wang J, Qin S, Liu D, Li Y and Lei W 2018 B/N co-doped carbon nanosphere frameworks as high-performance electrodes for supercapacitors *J. Mater. Chem.* A **6** 8053–8
[23] Lei W, Qin S, Liu D, Portehault D, Liu Z and Chen Y 2013 Large scale boron carbon nitride nanosheets with enhanced lithium storage capabilities *Chem. Commun.* **49** 352–4
[24] Zhang L *et al* 2020 MXene coupled with molybdenum dioxide nanoparticles as 2D-0D pseudocapacitive electrode for high performance flexible asymmetric micro-supercapacitors *J. Mater.* **6** 138–44
[25] Zhang L *et al* 2019 Shape-tailorable high-energy asymmetric micro-supercapacitors based on plasma reduced and nitrogen-doped graphene oxide and MoO2 nanoparticles *J. Mater. Chem.* A **7** 14328–36

[26] Huang P *et al* 2016 On-chip and freestanding elastic carbon films for micro-supercapacitors *Science (80–).* **351** 691–5
[27] Lin Y, Gao Y and Fan Z 2017 Printable fabrication of nanocoral-structured electrodes for high-performance flexible and planar supercapacitor with artistic design *Adv. Mater.* **29** 1–8
[28] Gao W, Singh N, Song L, Liu Z, Reddy A L M, Ci L, Vajtai R, Zhang Q, Wei B and Ajayan P M 2011 Direct laser writing of micro-supercapacitors on hydrated graphite oxide films *Nat. Nanotechnol.* **6** 496–500
[29] Li L, Lou Z, Han W, Chen D, Jiang K and Shen G 2017 Highly stretchable micro-supercapacitor arrays with hybrid MWCNT/PANI electrodes *Adv. Mater. Technol.* **2** 1600282
[30] Cao L *et al* 2013 Direct laser-patterned micro-supercapacitors from paintable MoS2 films *Small* **9** 2905–10
[31] Kim C, Kang D Y and Moon J H 2018 Full lithographic fabrication of boron-doped 3D porous carbon patterns for high volumetric energy density microsupercapacitors *Nano Energy* **53** 182–8
[32] Qin J, Wang S, Zhou F, Das P, Zheng S and Sun C 2019 Supercapacitors in water-in-salt gel electrolyte *Energy Storage Mater.* **18** 397–404
[33] Zhang P, Li Y, Wang G, Wang F, Yang S, Zhu F, Zhuang X, Schmidt O G and Feng X 2019 Zn-ion hybrid micro-supercapacitors with ultrahigh areal energy density and long-term durability *Adv. Mater.* **31** 1–6
[34] Zheng S, Ma J, Wu Z S, Zhou F, He Y B, Kang F, Cheng H M and Bao X 2018 All-solid-state flexible planar lithium ion micro-capacitors *Energy Environ. Sci.* **11** 2001–9
[35] Sun G, Yang H, Zhang G, Gao J, Jin X, Zhao Y, Jiang L and Qu L 2018 A capacity recoverable zinc-ion micro-supercapacitor *Energy Environ. Sci.* **11** 3367–74
[36] Xie Y, Zhang H, Huang H, Wang Z, Xu Z, Zhao H, Wang Y, Chen N and Yang W 2020 Nano energy high-voltage asymmetric MXene-based on-chip micro-supercapacitors *Nano Energy* **74** 1–9
[37] Kim J W, Augustyn V and Dunn B 2012 The effect of crystallinity on the rapid pseudocapacitive response of Nb2O5 *Adv. Energy Mater.* **2** 141–8
[38] Pech D, Brunet M, Durou H, Huang P, Mochalin V, Gogotsi Y, Taberna P L and Simon P 2010 Ultrahigh-power micrometre-sized supercapacitors based on onion-like carbon *Nat. Nanotechnol.* **5** 651–4
[39] Pantrangi M *et al* 2024 Flexible micro-supercapacitors: materials and architectures for smart integrated wearable and implantable devices *Energy Storage Mater.* **73** 103791
[40] Wu Z S, Parvez K, Feng X and Müllen K 2013 Graphene-based in-plane micro-supercapacitors with high power and energy densities *Nat. Commun.* **4** 2487
[41] Li H, Ma L, Han C, Wang Z, Liu Z, Tang Z and Zhi C 2019 Advanced rechargeable zinc-based batteries: recent progress and future perspectives *Nano Energy* **62** 550–87
[42] Song M, Tan H, Chao D and Fan H J 2018 Recent advances in Zn-ion batteries *Adv. Funct. Mater.* **28** 1–27
[43] Shao Y *et al* 2021 Regulating oxygen substituents with optimized redox activity in chemically reduced graphene oxide for aqueous Zn-ion hybrid capacitor *Adv. Funct. Mater.* **31** 1–11
[44] Cheng W, Fu J, Hu H and Ho D 2021 Interlayer structure engineering of MXene-based capacitor-type electrode for hybrid micro-supercapacitor toward battery-level energy density *Adv. Sci.* **8** 2100775

[45] Liu Z, Tian X, Xu X, He L, Yan M, Han C, Li Y, Yang W and Mai L 2017 Capacitance and voltage matching between MnO2 nanoflake cathode and Fe2O3 nanoparticle anode for high-performance asymmetric micro-supercapacitors *Nano Res.* **10** 2471–81

[46] Wang S, Wu Z S, Zhou F, Shi X, Zheng S, Qin J, Xiao H, Sun C and Bao X 2018 All-solid-state high-energy planar hybrid micro-supercapacitors based on 2D VN nanosheets and Co (OH)2 nanoflowers *Npj 2D Mater. Appl.* **2** 1–8

[47] Liu L, Lu Q, Yang S, Guo J, Tian Q, Yao W, Guo Z, Roy V A L and Wu W 2018 All-printed solid-state microsupercapacitors derived from self-template synthesis of Ag@PPy nanocomposites *Adv. Mater. Technol.* **3** 1–9

[48] Lin J, Peng Z, Liu Y, Ruiz-Zepeda F, Ye R, Samuel E L G, Yacaman M J, Yakobson B I and Tour J M 2014 Laser-induced porous graphene films from commercial polymers *Nat. Commun.* **5** 5–12

[49] Liu L, Ye D, Yu Y, Liu L and Wu Y 2017 Carbon-based flexible micro-supercapacitor fabrication via mask-free ambient micro-plasma-jet etching *Carbon* **111** 121–7

[50] Hyun W J, Secor E B, Kim C H, Hersam M C, Francis L F and Frisbie C D 2017 Scalable, self-aligned printing of flexible graphene micro-supercapacitors *Adv. Energy Mater.* **7** 1–8

[51] Zhang J, Zhang G, Zhou T and Sun S 2020 Recent developments of planar micro-supercapacitors: fabrication, properties, and applications *Adv. Funct. Mater.* **30** 1910000

[52] Lv H, Pan Q, Song Y, Liu X X and Liu T 2020 A review on nano-/microstructured materials constructed by electrochemical technologies for supercapacitors

[53] Wang R, Luo S, Xiao C, Chen Z, Li H, Asif M, Chan V, Liao K and Sun Y 2021 MXene-carbon nanotubes layer-by-layer assembly based on-chip micro-supercapacitor with improved capacitive performance *Electrochim. Acta* **386** 138420

[54] Park S H, Goodall G and Kim W S 2020 Perspective on 3D-designed micro-supercapacitors *Mater. Des.* **193** 108797

[55] Li W, Li Y, Su M, An B, Liu J, Su D, Li L, Li F and Song Y 2017 Printing assembly and structural regulation of graphene towards three-dimensional flexible micro-supercapacitors *J. Mater. Chem.* A **5** 16281–8

[56] Pal B, Yang S, Ramesh S, Thangadurai V and Jose R 2019 Electrolyte selection for supercapacitive devices: a critical review *Nanoscale Adv* **1** 3807–35

[57] Senthilkumar S T, Selvan R K, Melo J S and Sanjeeviraja C 2013 High performance solid-state electric double layer capacitor from redox mediated gel polymer electrolyte and renewable tamarind fruit shell derived porous carbon *ACS Appl. Mater. Interfaces* **5** 10541–50

[58] Dong L, Xu C, Yang Q, Fang J, Li Y and Kang F 2015 High-performance compressible supercapacitors based on functionally synergic multiscale carbon composite textiles *J. Mater. Chem.* A **3** 4729–37

[59] Hall P J, Mirzaeian M, Fletcher S I, Sillars F B, Rennie A J R, Shitta-Bey G O, Wilson G, Cruden A and Carter R 2010 Energy storage in electrochemical capacitors: designing functional materials to improve performance *Energy Environ. Sci.* **3** 1238–51

[60] Kim H and Popov B N 2003 A mathematical model of oxide/carbon composite electrode for supercapacitors *J. Electrochem. Soc.* **150** A1153

[61] Najib S and Erdem E 2019 Current progress achieved in novel materials for supercapacitor electrodes: mini review *Nanoscale Adv.* **1** 2817–27

[62] Behera R and Elanseralathan K 2022 A review on polyvinylidene fluoride polymer based nanocomposites for energy storage applications *J. Energy Storage.* **48** 103788

[63] Yassine M and Fabris D 2017 Performance of commercially available supercapacitors *Energies* **10** 1340

[64] Kumar N, Kumar A, Chandrasekaran S and Tseng T Y 2018 Synthesis of mesoporous NiFe2O4 nanoparticles for enhanced supercapacitive performance *J. Clean Energy Technol.* **6** 51–5

[65] Li Q, Horn M, Wang Y, MacLeod J, Motta N and Liu J 2019 A review of supercapacitors based on graphene and redox-active organic materials *Materials (Basel).* **12** 703

[66] Yibowei M E, Adekoya J G, Adediran A A and Adekomaya O 2021 Carbon-based nano-filler in polymeric composites for supercapacitor electrode materials: a review *Environ. Sci. Pollut. Res.* **28** 26269–79

[67] Iro Z S, Subramani C and Dash S S 2016 A brief review on electrode materials for supercapacitor *Int. J. Electrochem. Sci.* **11** 10628–43

[68] Lokhande P E, Chavan U S and Pandey A 2020 Materials and fabrication methods for electrochemical supercapacitors: overview *Electrochem. Energy Rev.* **3** 155–86

[69] Faid A Y and Ismail H 2019 Ternary mixed nickel cobalt iron oxide nanorods as a high-performance asymmetric supercapacitor electrode *Mater. Today Energy.* **13** 285–92

[70] Raza W, Ali F, Raza N, Luo Y, Kim K H, Yang J, Kumar S, Mehmood A and Kwon E E 2018 Recent advancements in supercapacitor technology *Nano Energy* **52** 441–73

[71] Yu L and Chen G Z 2020 Supercapatteries as High-performance electrochemical energy storage devices *Electrochem. Energy Rev.* **3** 271–85

[72] Wayu M 2021 Manganese oxide carbon-based nanocomposite in energy storage applications *Solids* **2** 232–48

[73] Yan J, Fan Z, Wei T, Qian W, Zhang M and Wei F 2010 Fast and reversible surface redox reaction of graphene-MnO2 composites as supercapacitor electrodes *Carbon N. Y.* **48** 3825–33

[74] Wang S, Ma J, Shi X, Zhu Y and Wu Z S 2022 Recent status and future perspectives of ultracompact and customizable micro-supercapacitors *Nano Res. Energy.* **1** 1–8

[75] Sung J H, Kim S J and Lee K H 2003 Fabrication of microcapacitors using conducting polymer microelectrodes , *J. Power Sources* **124** 343–50

[76] Jiang Q, Kurra N, Maleski K, Lei Y, Liang H, Zhang Y, Gogotsi Y and Alshareef H N 2019 On-chip MXene microsupercapacitors for AC-line filtering applications *Adv. Energy Mater.* **9** 1–9

[77] Beidaghi M and Wang C 2012 Micro-supercapacitors based on interdigital electrodes of reduced graphene oxide and carbon nanotube composites with ultrahigh power handling performance *Adv. Funct. Mater.* **22** 4501–10

[78] In J B, Hsia B, Yoo J H, Hyun S, Carraro C, Maboudian R and Grigoropoulos C P 2015 Facile fabrication of flexible all solid-state micro-supercapacitor by direct laser writing of porous carbon in polyimide *Carbon N. Y.* **83** 144–51

[79] Lobo D E, Banerjee P C, Easton C D and Majumder M 2015 Miniaturized supercapacitors: focused ion beam reduced graphene oxide supercapacitors with enhanced performance metrics *Adv. Energy Mater.* **5** 1500665

[80] Laszczyk K U, Kobashi K, Sakurai S, Sekiguchi A, Futaba D N, Yamada T and Hata K 2015 Lithographically integrated microsupercapacitors for compact, high performance, and designable energy circuits *Adv. Energy Mater.* **5** 1–6

[81] Ferris A, Reig B, Eddarir A, Pierson J F, Garbarino S, Guay D and Pech D 2017 Atypical properties of FIB-patterned RuOx nanosupercapacitors *ACS Energy Lett.* **2** 1734–9

[82] Shen D, Zou G, Liu L, Zhao W, Wu A, Duley W W and Zhou Y N 2018 Scalable high-performance ultraminiature graphene micro-supercapacitors by a hybrid technique combining direct writing and controllable microdroplet transfer *ACS Appl. Mater. Interfaces* **10** 5404–12

[83] Xu C, Jiang L, Li X, Li C, Shao C, Zuo P, Liang M, Qu L and Cui T 2020 Miniaturized high-performance metallic 1T-phase MoS2 micro-supercapacitors fabricated by temporally shaped femtosecond pulses *Nano Energy* **67** 104260

[84] Yuan Y *et al* 2020 Based micro-supercapacitors ultrafast fabrication *Nat. Commun.* **11** 6185

[85] Rogers J A, Someya T and Huang Y 2010 Materials and mechanics for stretchable electronics *Science* **327** 1603–7

[86] Choi C, Robert K, Whang G, Roussel P, Lethien C and Dunn B 2021 Photopatternable hydroxide ion electrolyte for solid-state micro-supercapacitors *Joule* **5** 2466–78

[87] Pei L, Zhuang P, Sun Y, Zhang X, Qiao H, Zhu K, Ajayan P M, Ye M and Shen J 2021 Nanosupercapacitors with fractal structures: searching designs to push the limit *J. Mater. Chem.* A **9** 17400–14

[88] Wen Z *et al* 2016 Self-powered textile for Wearable electronics by hybridizing fiber-shaped nanogenerators, solar cells, and supercapacitors *Sci. Adv.* **2** 1–8

[89] Li D, Yang S, Chen X, Lai W Y and Huang W 2021 3D Wearable fabric-based micro-supercapacitors with ultra-high areal capacitance *Adv. Funct. Mater.* **31** 1–7

[90] Pu X, Liu M, Li L, Han S, Li X, Jiang C, Du C, Luo J, Hu W and Wang Z L 2016 Wearable textile-based In-Plane microsupercapacitors *Adv. Energy Mater.* **6** 1601254

[91] Qiu M, Sun P, Cui G, Tong Y and Mai W 2019 A flexible microsupercapacitor with integral photocatalytic fuel cell for self-charging *ACS Nano* **13** 8246–55

IOP Publishing

Supercapacitors
Concepts and advances
Abha Misra

Chapter 8

Self-powered supercapacitors and batteries

8.1 Self-powered supercapacitors

The pursuit of miniaturization, multifunctionality, and wearability in electronic devices has led to a demand for compact, lightweight, flexible power modules. In addition, conventional energy storage devices such as batteries require frequent charging and also face some degradation. Thus, self-charging supercapacitors that overcome these problems are a potential solution. To combine energy storage and energy harvesting, the emergence of new and enhanced self-charging supercapacitors is crucial. Energy harvesting is a unique feature whereby a supercapacitor replenishes itself naturally by tapping into available energy sources such as light, vibration, and chemical and thermal energies. Some of the advantages of these supercapacitors are that they represent greener energy alternatives, offer long and reliable operation, and exhibit higher efficiency, allowing more compact designs for integration with wearable electronics, environmentally sensitive applications, and medical devices. The different parts of a self-powering supercapacitor are the energy harvesting part, the energy conversion module, and the energy storage unit. The energy harvesting module includes all the techniques and mechanisms that can tap into surrounding sources in order to charge the supercapacitor. The conversion part converts the harvested energy into a suitable form for high-power density storage that has reliable charging/discharging characteristics and a long cycle life.

8.2 Triboelectric nanogenerator supercapacitors

Here, a triboelectric nanogenerator (TENG) generates electricity. The overall device may comprise, for example, a conformal Ni layer and a reduced graphene oxide (rGO) film on the surface of common polyester yarn (rGO-Ni-yarn), a supercapacitor fabric for energy storage, and a wearable button-like sensor. The supercapacitor is made of two rGO-Ni-yarns arranged in parallel and a poly(vinyl alcohol) (PVA)/phosphoric acid (H_3PO_4) gel that functions as a solid electrolyte and a separator [1]. Another example is that of a metal–organic framework

doi:10.1088/978-0-7503-5944-3ch8

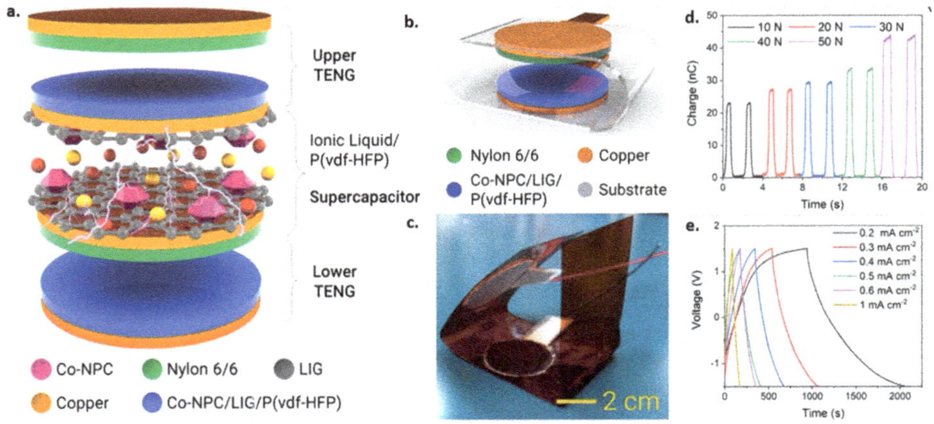

Figure 8.1. Schematic illustrations depicting (a) the fabrication of a MOF-derived cobalt-based nanoporous Co-NPC/LIG/Cu-based SPC electrode and (b) the positive and negative triboelectric layers of a TENG. (c) Optical image of a fabricated TENG. (d) Charges generated in a TENG under different compressive forces. (e) The galvanostatic charge/discharge response of an SPC at current densities ranging from 0.2 to 1 mA cm^{-2}. Adapted from [2], Copyright (2022) with permission from Elsevier.

(MOF)-derived cobalt-based nanoporous carbon/laser-induced graphene/copper (Co-NPC/LIG/Cu)-based self-powered capacitor (SPC) electrode [2]. Here, mechanical motion is converted into electrical energy using triboelectrification and electrostatic induction. When two materials come into contact and separate, they acquire opposite charges. This phenomenon is known as triboelectrification. As the charged materials separate, an electric field is created, which induces opposite charges in the electrodes attached to the materials [3]. No external circuitry is required for triboelectrification relying solely on triboelectrochemical mechanisms [2]. An example of such a structure with positive and negative triboelectric layers is shown in figures 8.1(a)–(c). Figure 8.1(d) shows the charge generated under different compressive forces and the charge–discharge characteristics (figure 8.1(e)). Figure 8.2 shows the simplified working principle of the self-powering supercapacitor; it also demonstrates that the amount of voltage generated increases with an increase in the compressive force applied across the two faces.

8.3 Piezoelectric supercapacitors

Piezoelectric methods convert mechanical energy (pressure, vibration) into electricity. Various techniques involve transforming mechanical energy into electrical energy using either external integration methods (EIMs) or internal integration methods (IIMs). The latter are commonly used in wearable devices, where body movement can charge the supercapacitor because of their integrated and compact setups. Materials that have high piezoelectric constants, high permittivity, flexibility, and ferroelectric properties are generally preferred due to their piezoelectric performance. Some materials that have these properties are lead zirconate titanate

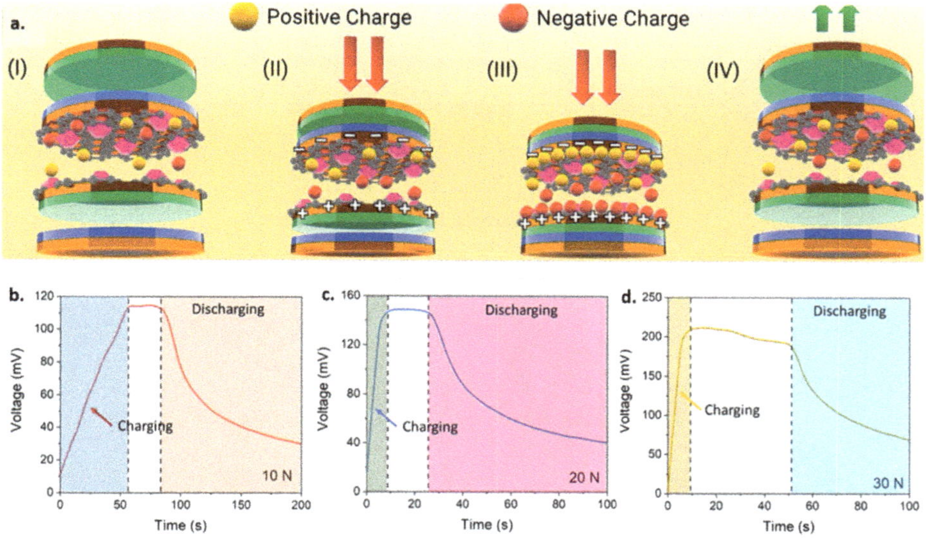

Figure 8.2. (a) Schematic showing the working principle of a self-charging SPC. (b)–(d) Self-charging properties of the SPC under various applied forces of 10, 20, and 30 N. Adapted from [2], Copyright (2022) with permission from Elsevier.

(PZT) [4], barium titanate ($BaTiO_3$), zinc oxide (ZnO) [5], polyvinylidene difluoride (PVDF) [6], poly(vinylidene fluoride-trifluoroethylene) (P(VDF-TrFE)) [7], and polydimethylsiloxane (PDMS) [8]. An example of use of such a material in a supercapacitor is (PVDF)-ZnO [5]. A self-charging supercapacitor power cell (SCSPC) has been developed that comprises (PVDF)-ZnO as a piezoelectric material as well as a separator, PVA-H_3PO_4 as a gel electrolyte, and manganese oxide (MnO_2) nanowires as positive and negative electrodes [5].

Here, the compressive stress on the structure causes the supercapacitor to self-charge. Combining different nanomaterials, such as ZnO and Mo–Fe doped MnO_2, has shown promise in enhancing the performance of self-charging supercapacitors [5]. Integrating piezoelectric materials such as PVDF into supercapacitors allows for efficient energy harvesting from mechanical stress [9]. Figure 8.3(a) shows a schematic diagram of the fabricated SCSPC, including the charge transfer mechanism under compression and the different intermediate layers representing the various charges. Figure 8.3(b) shows the cyclic voltammetry (CV) of the supercapacitor at different scan rates, and figure 8.3(c) shows the charging and discharging mechanisms under compression and in the absence of compression, respectively.

A heterostructure made of ZnO–NiO has been shown to form a strain-dependent charge layer. The inherent piezoelectric characteristic of ZnO causes the creation of a piezo potential and, consequently, a self-powering response when a horizontal tensile strain is applied (figure 8.4(a)). However, p-NiO redesigns this intrinsic piezo potential to lower the interfacial leakage current. The end effect is an improved piezo potential, which raises the heterostructure's self-powering response [10]. The same

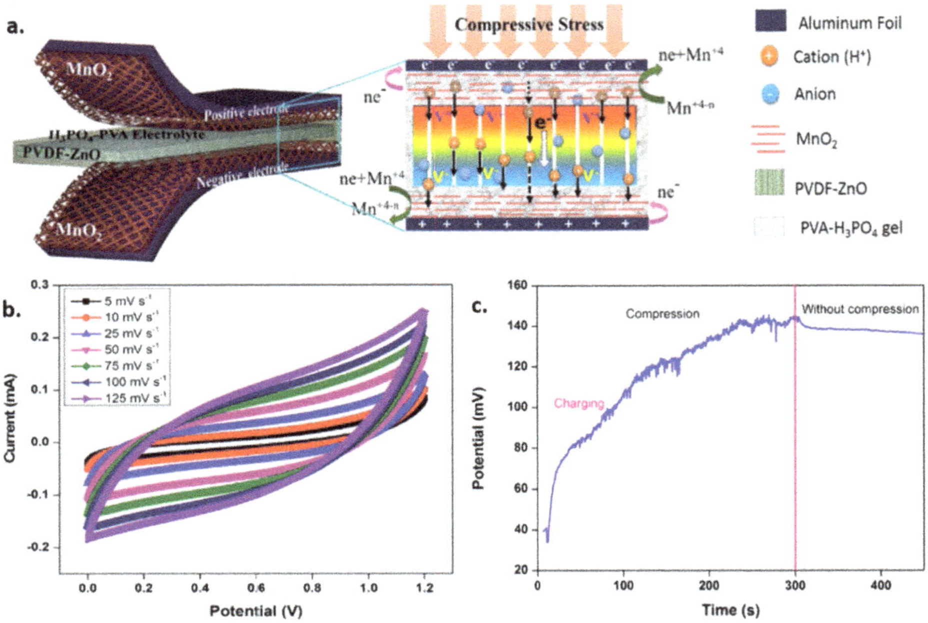

Figure 8.3. (a) Schematic illustration of the fabricated SCSPC. MnO_2 coated on aluminum foil is used for the positive and negative electrodes, and PVDF-ZnO film is used as a separator. (b) CV scans of the SCSPC at different scan rates varying from 5 to 125 mV s^{-1}. (c) Self-powering behavior in an SCSPC due to impact by a human palm. The fact that the voltage continues to increase during vibration indicates that the charging process is due to ion migration. Reprinted with permission from [5]. Copyright (2015) American Chemical Society.

properties are often undesirable in certain cases, such as when we need reliable capacitive performance irrespective of the stress or strain conditions in a structure. As a result, there have been demonstrations of structure made in the form of zigzag edge electrodes and prepared on a flexible poly(ethylene terephthalate) (PET) substrate, which are immune to such variations in stress and strain. Commercially available glass substrates were used to build carbon fiber (CF)–MoS_2-based zigzag edge and planar edge microsupercapacitors. Microsupercapacitors with flat and zigzag edges were created using prepatterned and specially produced stainless steel masks. The process of creating a microsupercapacitor began with the cleaning of the glass substrate and stainless-steel masks in an ultrasonic bath for five minutes each, using deionized (DI) water, isopropyl alcohol (IPA), and ethanol. A nitrogen gas blower was then used to dry the surfaces. Eight in-plane electrode fingers, four for each polarity, comprised the planar edge interdigitated mask. Each finger measured 1×12.6 mm^2, with a 0.7 mm space between adjacent fingers. Every electrode was connected to a 0.5×0.5 cm^2 contact pad. The zigzag edge interdigitated mask also had eight in-plane electrode fingers, with four fingers per polarity. Each finger's diameter was consistently between 0.2 and 1 mm, with a 0.7 mm gap between neighboring fingers. Figure 8.4(b) illustrates the device's mechanical flexibility under various bending states, and CV measurements were taken at a scan rate of 70 mV s^{-1}

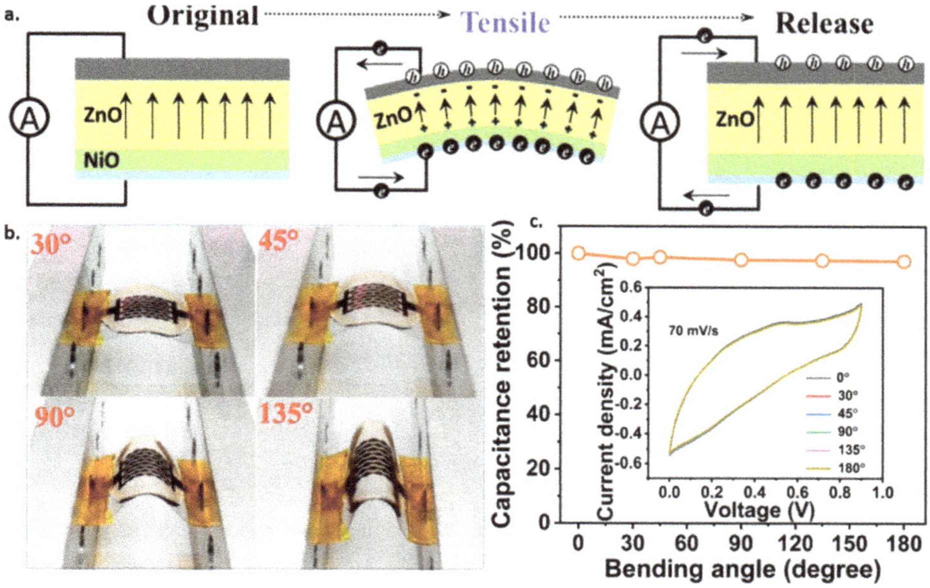

Figure 8.4. (a) Self-powered dynamic strain monitoring in a heterostructure (HS) with NiO for piezo-polarized charge generation in ZnO. Reproduced from [10] with permission from the Royal Society of Chemistry. (b) Photographs of the microsupercapacitor with a zigzag edge electrode on a PET substrate at different bending angles. (c) Capacitance retentions at different bending angles, Inset: CV responses of the microsupercapacitor with a zigzag edge electrode at different bending angles. Parts (b) and (c) reproduced from [11] with permission from the Royal Society of Chemistry.

to gauge the associated electrochemical performance. The CV responses demonstrate the device's electrochemical stability on the PET substrate, which exhibits no bending-induced alterations (inset of figure 8.4(c)). Additionally, figure 8.4(c) illustrates the link between capacitance retention and bending angle; after the device was bent by 180°, it still retained 99% of its initial capacitance [11].

8.4 Thermoelectric supercapacitors

In thermoelectric (TE) supercapacitors, ionic thermoelectric (iTE) materials cause ion migration under temperature gradients due to the Soret effect [12]. Electrolytes such as NaOH–polyethylene oxide (PEO) have a high Seebeck coefficient of 11.1 mV K^{-1} and show superior TE performance compared to that of traditional electronic TE materials. However, NaOH–PEO has poor moldability, is difficult to encapsulate, and leaks easily. Thus, film-like ionogels have attracted significant attention. These are usually based on polymers such as poly(vinylidene fluoride-co-hexafluoropropylene) (PVDF-HFP) and poly(3,4-ethylenedioxythiophene) polystyrene sulfonate (PEDOT: PSS), which are doped with functional ions from small molecules or ionic liquids to build ion-selective networks to improve the iTE properties. This strategy allows the Seebeck coefficient of an iTE material to reach 26.1 mV K^{-1} [13]. A stretchable bifunctional iTE supercapacitor has also been demonstrated based on a stretchable and environmentally friendly NaCl–polyacrylamide/sodium carboxymethyl cellulose

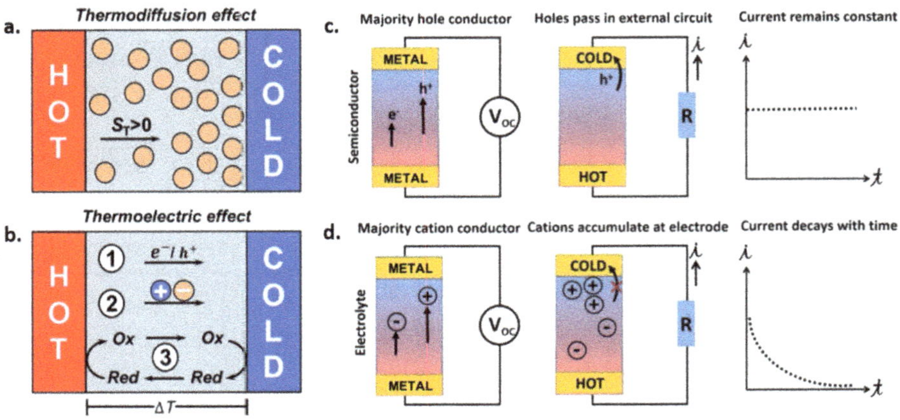

Figure 8.5. Schematic showing (a) the thermodiffusion (ST>0) of particles and (b) three different approaches used to obtain TE effects: (1) electronic conductors, (2) non-redox electrolytes, and (3) redox electrolytes. (c) TE effects produced in a semiconductor by electronic charge carriers (electrons or holes) and (d) in an electrolyte by ionic charge carriers. Adapted from [18]. CC BY 4.0.

(NaCl–PMSC) electrolyte [14] and carbon nanotube–polyacrylamide (CNT–PAM) electrodes. Both the gel electrolyte and the electrodes were fabricated using a synergistic cross-linking approach. Adjustment of the Na^+ content in the ionogel allowed the TE ionogel electrolyte to exhibit a maximum Seebeck coefficient of 17.1 mV K^{-1} [14]. The ions propagate as electronic charge carriers depending on the temperature gradient. They are eventually trapped at the electrode–electrolyte junction, forming an electric double layer (EDL) [15], as shown in figure 8.5. This is even more pronounced when conductive polymers and carbon nanotubes (CNTs) are used. Thus, modulating the thermal energy using the iTE method enhances the electrical performance of the device. The maximum electrical energy is given by $E_{ch} = \frac{1}{2}CV^2_{\text{thermo}} = \frac{1}{2}C(\Delta TS_I)^2$ [16]. This quadratically increases with the Seebeck coefficient [17] for these materials.

A thermogalvanic cell consists of two electrodes at varying temperatures in contact with an electrolyte that contains a redox pair. The oxidant (ox) and reductant (red) of a single electrochemical half-reaction are present in the redox electrolyte. Crucially, this half-reaction's electron transport at the electrode needs to be reversible. The Soret (or Ludwig–Soret) effect, often referred to as Soret diffusion [12], is the process by which distinct species react differently to the same temperature gradient when thermodiffusion takes place in a fluid mixture of two or more different constituents. This results in molecular separation within the mixture [17]. Stability, output voltage, efficiency, and performance are factors that can improve the output. The temperature-dependent electric conductivity, thermal expansion, and electronic structure may all reveal the dynamics of TE charge storage. Notably, unlike the case of porous electrodes, the actual process of thermal charging does not preclude the occurrence of either irreversible or reversible temperature-dependent charge transport at the electrode–electrolyte interface [19]. The main element that should be examined to precisely assess the effectiveness of thermal charging is the pore structure. Because they have distinct surface structures, nanotubes, carbides, and carbon might be suitable starting points, as they reveal more about their pore structure than materials with irregular or large pores.

8.5 Solar energy or light-based supercapacitors

It is essential to achieve high energy density in supercapacitors to improve their cycle life and output efficiency. In this respect, the piezoelectric and TE options for self-charging solve this problem; however, they are relatively slow. There is therefore a requirement for self-charging methods that are both efficient and fast. The solar or light-based self-charging phenomenon occurs under illumination and involves the self-storage of photogenerated electrical signals [20]. There are several different ways of harnessing the optical self-charging capabilities of supercapacitors. Some of them involve the use of solar cells [21], while others involve composite materials that directly harness a wide spectral range to enable effective charging under illumination. One of these methods involves the use of polyaniline (PANI) composites with CNTs. The photogenerated current density of these materials is about 2 mA g^{-1}. One example of a supercapacitor made from these materials consists of two slightly separated electrodes that are well solidified in a PVA gel electrolyte containing H_2SO_4. These electrodes consist of PANI layer on indium tin oxide (ITO) glass (PANI@ITO) and a composite film with PANI coated on a CNT film (PANI/CNT) as templates [22]. This yields a transparent supercapacitor with PANI as both the capacitive electrode material and the photoactive layer, with a specific capacitance of 140 F g^{-1}. Another method is to use a few layers of graphene–MXene ($Ti_3C_2T_x$) quantum dots (QDs) on fluoride-doped tin oxide (FTO). Here, the graphene reduces charge scattering and is also sensitive to the UV spectrum [23], while the MXene facilitates enhanced light interaction due to its wide bandgap [24].

Figure 8.6 shows the excitation of this device in the UV range, the charge storage after the generation of an EDL upon UV excitation, and the corresponding current density at 365 nm. It also shows the significant improvement of the areal capacitance due to the hybrid MXene/graphene structure. Another technique is to hybridize two different materials that are sensitive to different parts of the spectrum to achieve broadband excitation and charging of the supercapacitor. For example, ZnO interacts with the UV part of the spectrum, and CdS interacts with the visible part of the spectrum. Therefore, a structure with CdS nanoparticles grown on ZnO nanorods has a broadband response to incident light.

Figure 8.7 shows the enhanced response of the photochargeable supercapacitor, which exhibits a broadband response to incident interacting light. With a remarkable increase in areal capacitance compared to its dark value ranging between approximately 238% and 1264% across a wide spectral range (λ ~365–520 nm), the symmetric supercapacitor (SSC) created using this heterostructure transforms light energy into electrochemical energy under illumination [25].

Recently, a novel synthesis method based on ZnO nanorods (figure 8.8(a)), which led to highly porous electrodes, showed a record capacitance enhancement under

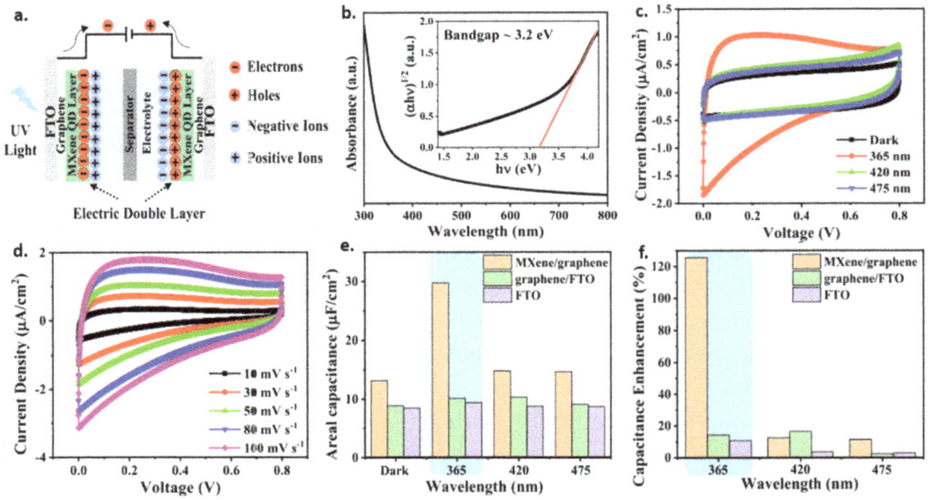

Figure 8.6. (a) Schematic illustrating the charge storage mechanism of an MXene QD/graphene/FTO device under UV illumination. (b) Optical absorption spectrum of MXene QDs and (inset) the corresponding Tauc plot. (c) MXene QD/graphene/FTO supercapacitors in the dark and under illumination at 365, 420 and 475 nm at a scan rate of 50 mV s^{-1}. (d) CV curves of the MXene QD/graphene/FTO device under 365 nm illumination at different scan rates from 10 to 100 mV s^{-1}. (e) The areal capacitance and (f) the enhancement in capacitance observed in the CV responses of the devices at a scan rate of 50 mV s^{-1} under different wavelengths of illumination and dark conditions. Adapted from [24], Copyright (2024) with permission from the American Chemical Society.

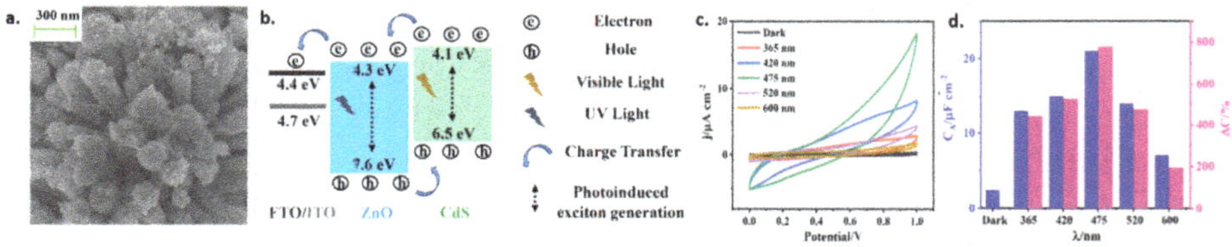

Figure 8.7. (a) Field emission scanning electron microscopy (FE-SEM) image showing the microstructure of ZnO/CdS nanorods grown on a substrate. (b) Schematic showing the charge transfer mechanism at the interface of a ZnO/CdS heterojunction. (c) CV responses of FTO/ZnO/CdS nanorod devices in dark and illuminated conditions at a scan rate of 50 mVs^{-1}. (d) Measured capacitance, C_A, and ΔC under dark and illuminated conditions (λ ~365, 420, 475, 520, and 600 nm) at a scan rate of 50 mVs^{-1}. Adapted from [25], Copyright (2024) with permission from Elsevier.

UV illumination, achieving an increment of ~3006% (an improvement of more than ten times compared to previous records). Here, two identical ZnO/FTO electrodes were used to create an SSC. The effects of UV light on the device's energy storage capacity were examined using two electrolytes, namely BMIM–BF4 (an ionic liquid) and PVA/KCl (a gel electrolyte). An electrochemical study of the ZnO/FTO SC with a PVA/KCl gel electrolyte under dark conditions and without UV light is displayed in figure 8.8(b); from 10 to 100 mV s^{-1}, the CV curves in figure 8.8(b) are measured

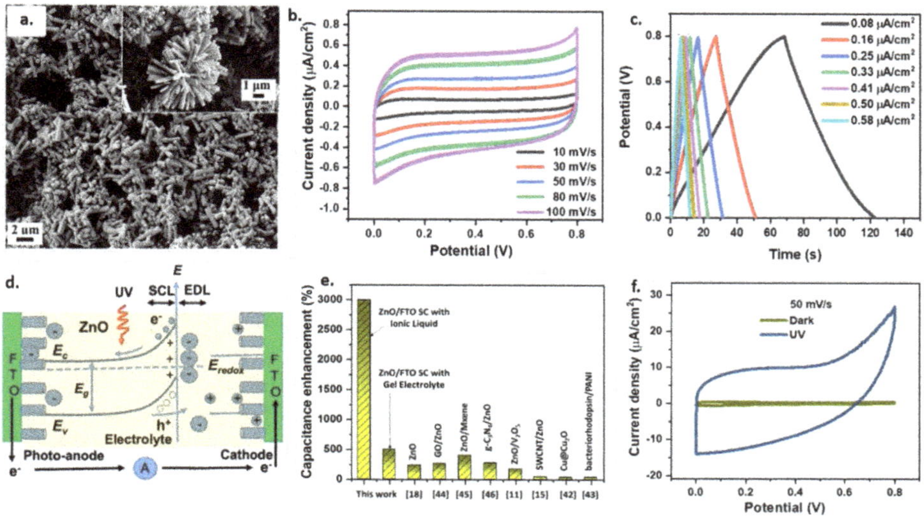

Figure 8.8. (a) Microstructure of ZnO nanorods. (b) Electrochemical responses of a ZnO/FTO SC with a PVA/KCl gel electrolyte: CV and (c) galvanostatic charge/discharge (GCD) analyses at different scan rates and current densities. (d) Illustration of the interfaces between the optically responsive anode and cathode and the electrolyte. (e) Percentage capacitance enhancement compared with other studies. (f) CV analysis at a scan rate of 50 mVs^{-1} of the ZnO/FTO supercapacitor with a 1-butyl-3-methylimidazolium tetrafluoroborate (BMIM–BF_4) electrolyte under dark and UV-illuminated conditions. Adapted from [26] with permission from the Royal Society of Chemistry.

at different scan rates. As illustrated in figure 8.8(c), the charge–discharge characteristics are observed at current densities ranging from 0.08 to 0.58 $mA\,cm^{-2}$. Compared to those of a solid-state supercapacitor, the electrochemical impedance spectroscopy (EIS) measurements of this device also show intriguing changes in dark and light settings. The plots shift from linear to curved at high frequencies, indicating the presence of charge transfer resistance (R_{ct}). Exposure to UV light may cause a modest increase in R_{ct} that results from the resistance of the solution. Easy ion transport in the ionic liquid electrolyte is indicated by the fact that the CV curves of the electrolyte have no peak, even under illumination. Moreover, increased slopes at higher voltages correlate with higher currents at higher voltages (see the CV curves in figures 8.8(e) and (f)). The operation of a photorechargeable supercapacitor is depicted in figure 8.8(d). The greater charge mobility of liquid electrolyte supercapacitors compared to gel electrolytes explains their larger energy storage capacity. The band bending at equilibrium causes a space charge layer (SCL) to form at the interface between the ZnO nanorods and the electrolyte when exposed to UV light. An EDL forms at the interface because photogenerated electron–hole pairs are separated by the inherent potential. Increased ionic flux through the porous ZnO nanorods raises the charge storage capacity because the enhanced charge separation brought about by UV light also strengthens the electrostatic attraction between the electrodes and oppositely charged ions in the electrolyte. Under UV light, the study showed excellent results with the liquid electrolyte and an electrochemical capacitance improvement of almost 3006%. It was demonstrated that the diffusion-controlled capacitance contribution rose under UV illumination [26].

8.6 Moisture-based self-charging in supercapacitors

This approach uses materials that effectively collect moisture from the ambient air, which subsequently reacts with the materials to produce ions. A potential difference is produced by these ions' participation in electrochemical reactions inside the supercapacitor. An example of a moisture-induced self-charging device (MISD) was created by combining two CNT electrodes with HCl/PVA electrolyte gel and $CaCl_2$ on one side [27]. However, converting moisture into usable electricity is still relatively inefficient and unreliable, as the device's performance is heavily dependent on the humidity of the environment [28]. Moisture-enabled electric generation (MEG) uses water molecules present in natural and biological settings and converts the corresponding chemical, mechanical, or thermal energy into its electrical counterpart within the structural part of the device [29]. An overall MEG system has two parts: the functional layers and the electrode layers. MEG devices are generally categorized into two categories: one based on ion separation and the other based on the results of redox reactions; however, there are cases where these occur simultaneously. Ions can be directionally diffused to the opposite side of the device to induce electricity generation under the influence of a concentration gradient and the Debye shielding effect (inside a charged nanochannel, if the channel's diameter is less than twice its Debye length, counter-ions pass through the nanochannel while homo-ions are blocked at the channel entrance, achieving an ion-shielding effect) [30].

The moisture/water molecules come into contact with the accumulated charge on the solid material. These accumulated charges form an EDL by attracting the polar molecules of water. This layer leads to confinement and charge transfer and also creates a potential difference. Directional ion migration is made possible by the introduction of active electrodes, which cause redox reactions to consume migrating ions.

Figure 8.9 shows porous fluorinated oxidized graphene (FOG) and oxidized graphene (OG). The ionic concentration differential in the heterojunction is further

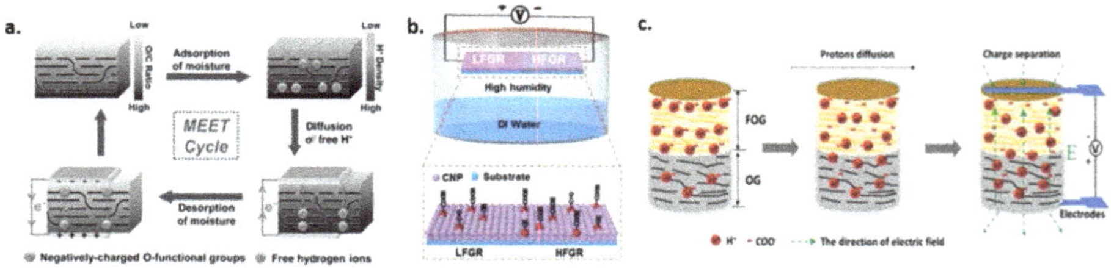

Figure 8.9. (a) Process of moisture-electric annealing. (b) Porous carbon film device used for harvesting adsorption-induced potential. (c) Schematic showing the mechanism responsible for the spontaneous diffusion of H^+ and charge separation in a H-FOG device. Reproduced from [31].

amplified by the added fluorine, which speeds up the ionization of the carboxyl groups and causes directional ion migration and power generation [31]. The duration of generation is one of MEG's primary performance characteristics, and it has a direct impact on the device's capability for subsequent applications. Limited functional groups and a modest concentration gradient in early MEG designs led to short power production times. The power generation duration has evolved from the initial tens of seconds to months, while functional materials with a high number of functional groups and the ability to maintain an ion gradient have emerged with the ongoing development of MEGs.

8.7 Optically sensitive electrochemical reactions

The basis of this approach is special electrode materials that can undergo reversible redox reactions. These materials typically exhibit pseudocapacitance, allowing for additional energy storage beyond that provided by the electrostatic double layer. A suitable electrolyte is chosen to facilitate the electrochemical reactions. It must be compatible with both the electrode materials and the desired redox reactions. The system requires an external energy source, such as light, heat, or mechanical energy, to trigger the electrochemical reactions. With a layered structure of graphene–Ag conductive epoxy–graphene foam, the graphene foam SC shows excellent electrochemical and supercapacitive performance. At a current density of 0.67 mA cm^{-2}, it has an areal capacitance (38 mF cm^{-2}) that is about three times higher than the values reported for flexible carbon-based SCs [32]. In another study, siloxene sheets were prepared via the topochemical deintercalation of calcium from $CaSi_2$ in the presence of an ice-cold HCl solution. Siloxene/PVDF piezofibers were electrospun throughout the fabrication process. Using electrospun siloxene–PVDF piezofibers filled with ionogel electrolyte as the separator and siloxene sheet-coated carbon cloth as two symmetric electrodes, a siloxene SCSPC device was created [33]. Material-based self-powered solid-state supercapacitors (SPSCs) provide continuous energy storage [34]. The maximum self-charging capacitance for a supercapacitor has been reported for a core–shell structured peptide-Co_9S_8 supercapacitor with a fairly high specific capacitance of 1300 mF cm^{-2} (1800 F g^{-1}) at 0.7 mA cm^{-2}. Meanwhile, the supercapacitor is self-powered by a TENG [35].

One of the highest recorded specific capacitances for a supercapacitor has been reported for a supercapacitor electrode composed of a well-aligned cobalt monoxide (CoO) nanowire array grown on 3D nickel foam with polypyrrole (PPy). The specific capacitance of this arrangement was reported to be 2223 F g^{-1}, with 99.8% capacitance retention after 2000 cycles [36]. Another arrangement of two $Ni(OH)_2$ layers with different morphologies and the same rGO layer between them, created through a facile one-step hydrothermal approach, formed a composite electrode that showed a maximum specific capacitance of 3328.7 F g^{-1} at a current density of 7 mA cm^{-2} (1.5 A g^{-1}) in a 1 M KOH solution under favorable conditions [37]. Functionalized CNTs such as MnO_2/CNT have shown capacitances of up to 1370 F g^{-1} [38]. Methods for synthesizing MnO_2-functionalized CNTs using a low-temperature coprecipitation method [39] promise extensive increases in self-capacitance capabilities.

To create a small and intelligent self-charging power package, an electrochromic microsupercapacitor with superior electrochemical qualities was combined with a hybrid piezo/triboelectric nanogenerator with high output power (see figure 8.10). The two main parts of the package are an arch-shaped nanogenerator and an microsupercapacitor array based on an Ag nanowire/NiO active electrode material [40]. At various charging stages, it allowed for stable and reversible chromatic shifts while providing adequate real-time information about its operational state [40].

More recently, measurements have shown that it is possible to achieve electrochemical capacitances as high as 55 mF cm^{-2} using a MoS_2–graphene symmetric field-effect transistor (FET) device with a gel electrolyte made from PVA/H_2SO_4. The output characteristics of the device show the presence of a Schottky contact between the MoS_2 layer and the metal. In addition, this device shows symmetric behavior under forward and reverse biases.

Figure 8.11 shows the fabricated MoS_2/Gr-based FET, in which the heterostructure functions as the channel. Confirming the pattern derived from the capacitance enhancement computed from the CV curves at various V_{gs} values, the greatest enhancement (of about 1677%) was seen at a V_{gs} value of −25 V. The active layers of MoS_2 and Gr have respective thicknesses of 61.9 and 6.2 nm at the cathode (C) and 7.6 and 7.8 nm at the anode (A). The symmetric electrodes of 2D MoS_2 and

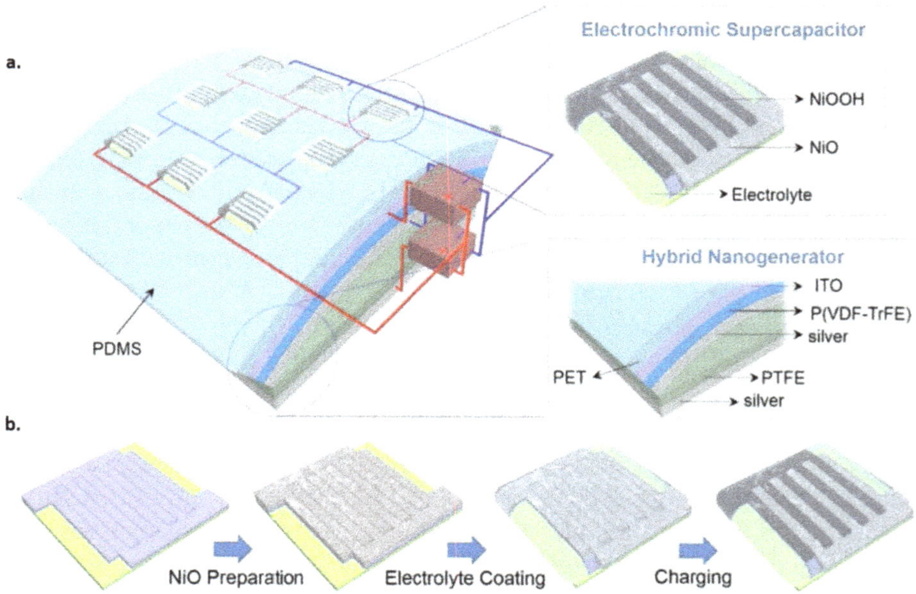

Figure 8.10. (a) Schematic showing the self-charging power package. The top inset shows the structure of the electrochromic microsupercapacitor; the bottom inset shows the structure of the hybrid nanogenerator. (b) Fabrication and electrochromic process of the electrochromic microsupercapacitor [40] John Wiley & Sons. © 2018 WILEY-VCH Verlag GmbH & Co. KGaA, Weinheim

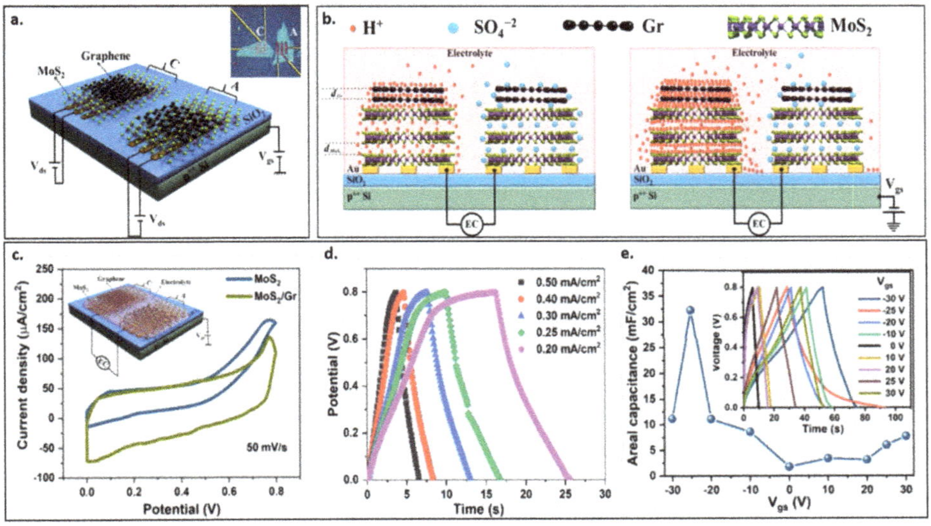

Figure 8.11. (a) Schematic diagrams of the cathode (C) and anode (A) of the FET device. (b) Schematic representation of the charge storage mechanism of the fabricated nanoelectrochemical capacitor with and without the application of V_{gs} (right and left). (c) CV analysis of MoS_2 and MoS_2/Gr devices at a scan rate of 50 mV s^{-1}. (inset) Schematic of the electrochemical capacitor. (d) GCD curves of the MoS_2/Gr device at different current densities. (e) Areal capacitance calculated from GCD measurements at 0.4 mA cm^{-2} and at different values of V_{gs} with (inset) the corresponding GCD curves. Adapted from [41], Copyright (2023) with permission from the American Chemical Society.

graphene flakes forming this electrochemical ultramicrocapacitor are directly transferred onto the metal current collectors with the gel electrolyte in between them at a separation of a few micrometers to form the whole system. By electrostatically doping MoS_2, few-layer MoS_2 can be used to create an electric-field-controlled energy storage device. This is further linked to improved charge storage through electric double-layer capacitance. By correcting issues encountered with liquid electrolytes, solid-state gel-based electrolytes make it possible to incorporate an electrochemical ultramicrocapacitor device into any electronic system. Compared to MoS_2 alone, our MoS_2/Gr-based electrochemical ultramicrocapacitor showed an exceptionally high electrochemical capacitance enhancement. In layered MoS_2 and graphene, gate-tunable charge carriers help to increase the electrochemical capacitance by approximately 3000% at −25 V of the back-gate voltage (V_{gs}). Higher charge mobility and an increase in the number of charge storage sites are responsible for the capacitance improvement. Galvanostatic charge–discharge (GCD) analyses were also carried out at various current densities ranging from 0.2 to 0.5 mA cm^{-2}. The observed GCD characteristics are displayed in the inset of figure 8.11(e), showing the charge–discharge behavior of the device with a back gate. Applying V_{gs} changes the GCD curve. For negative values of V_{gs}, the device's charging and discharging times increase significantly [41].

Electrochemical SCs are emerging as a potentially useful energy storage technology. High-performance SC electrodes that simultaneously guarantee high capacitance, cyclic stability, and exceptional rates require more research. This chapter has also discussed various new materials and technologies for SCs and the latest advancements in SC electrode materials. To enable these materials and innovative gadgets to satisfy rising energy demands, ongoing research is required. Even though this sector has seen some positive outcomes, the development of a new generation of SCs is still in its early stages [42].

References

[1] Pu X *et al* 2016 Wearable self-charging power textile based on flexible yarn supercapacitors and fabric nanogenerators *Adv. Mater.* **28** 98–105

[2] Shrestha K *et al* 2022 A triboelectric driven rectification free self-charging supercapacitor for smart IoT applications *Nano Energy* **102** 107713

[3] Wang J *et al* 2016 All-plastic-materials based self-charging power system composed of triboelectric nanogenerators and supercapacitors *Adv. Funct. Mater.* **26** 1070–6

[4] Zhang Y, Zhang Y, Xue X, Cui C and He B 2014 PVDF–PZT nanocomposite film based self-charging power cell *Nanotechnology* **25** 105401–8

[5] Ramadoss A, Saravanakumar B, Lee S W, Kim Y-S, Kim S J and Wang Z L 2015 Piezoelectric-driven self-charging supercapacitor power cell *ACS Nano* **9** 4337–45

[6] Song Y C R, Jin H, Li X, Fei L, Zhao Y, Huang H, Lai-Wa Chan H and Wang Y 2015 A rectification-free piezo-supercapacitor with a polyvinylidene fluoride separator and functionalized carbon cloth electrodes *J. Mater. Chem.* A **3** 14963–70

[7] Lu Y, Jiang Y, Lou Z, Shi R, Chen D and Shen G 2020 Wearable supercapacitor self-charged by P(VDF-TrFE) piezoelectric separator *Prog. Nat. Sci. Mater. Int.* **30** 174–9

[8] Okonkwo P C, Collins E and Okonkwo E 2017 *Application of Biopolymer Composites in Super Capacitor* (Amsterdam: Elsevier)

[9] Sahoo S, Krishnamoorthy K, Pazhamalai P, Mariappan V K, Manoharan S and Kim S J 2019 High performance self-charging supercapacitors using a porous PVDF-ionic liquid electrolyte sandwiched between two-dimensional graphene electrodes *J. Mater. Chem.* A **7** 21693–703

[10] Nandi S, Kumar S and Misra A 2021 Zinc oxide heterostructures: advances in devices from self-powered photodetectors to self-charging supercapacitors *Mater. Adv.* **2** 6768–99

[11] Kumar S, Mukherjee A, Telpande S, Mahapatra A D, Kumar P and Misra A 2023 Role of the electrode-edge in optically sensitive three-dimensional carbon foam-MoS_2 based high-performance micro-supercapacitors *J. Mater. Chem.* A **11** 4963–76

[12] Zhao D *et al* 2016 Ionic thermoelectric supercapacitors *Energy Environ. Sci.* **9** 1450–7

[13] Kim B, Hwang J U and Kim E 2020 Chloride transport in conductive polymer films for an n-type thermoelectric platform *Energy Environ. Sci.* **13** 859–67

[14] Yang X *et al* 2022 High-performance ionic thermoelectric supercapacitor for integrated energy conversion-storage *Energy Environ. Mater.* **5** 954–61

[15] Bonetti M *et al* 2015 Thermoelectric energy recovery at ionic-liquid/electrode interface *J. Chem. Phys.* **142** 244708

[16] Verma S, Padha B, Ahmed A, Singh R, Dubal D P and Arya S 2024 Thermoelectric supercapacitors: materials, challenges and future outlook *Prog. Energy* **6** 042002

[17] Al-Zubaidi A, Ji X and Yu J 2017 Thermal charging of supercapacitors: a perspective *Sustain. Energy Fuels* **1** 1457–74

[18] Massetti M *et al* 2021 Unconventional thermoelectric materials for energy harvesting and sensing applications *Chem. Rev.* **121** 12465–547
[19] Wang H *et al* 2017 Ionic thermoelectric figure of merit for charging of supercapacitors *Adv. Electron. Mater.* **3** 1700013
[20] Rani J R *et al* 2023 Realizing high energy density supercapacitors assisted by light-induced charging *J. Power Sources* **576** 233197
[21] Liu R *et al* 2017 Silicon nanowire/polymer hybrid solar cell-supercapacitor: a self-charging power unit with a total efficiency of 10.5% *Nano Lett.* **17** 4240–7
[22] Yin Y, Feng K, Liu C and Fan S 2015 A polymer supercapacitor capable of self-charging under light illumination *J. Phys. Chem.* C **119** 8488–91
[23] Sutradhar A, V P K, Das Mahapatra A, Sharma P and Misra A 2024 Effect of oxygen on ammonia sensing of large area graphene *New J. Chem.* **48** 15338–45
[24] Das Mahapatra A, Kumar S, Chauhan P S, Mondal A, Sutradhar A and Misra A 2024 Large area graphene-MXene quantum dot based high performance photochargeable supercapacitor *ACS Appl. Energy Mater.* **7** 5801–10
[25] Mahapatra A D, Kumar S, Sutradhar A, Sahoo S and Misra A 2024 ZnO/CdS based high performance broadband photo-chargeable flexible supercapacitor *Electrochim. Acta* **474** 143507
[26] Chauhan P S *et al* 2024 Influence of electrolyte on the photo-charging capability of a ZnO-FTO supercapacitor *J. Mater. Chem.* A **12** 22725–36
[27] Luo Z, Liu C and Fan S 2019 A moisture induced self-charging device for energy harvesting and storage *Nano Energy* **60** 371–6
[28] Wang L *et al* 2024 Moisture-enabled self-charging and voltage stabilizing supercapacitor *Nat. Commun.* **15** 1–9
[29] Liu X, Gao H, Sun L and Yao J 2024 Generic air-gen effect in nanoporous materials for sustainable energy harvesting from air humidity *Adv. Mater.* **36** 1–10
[30] Van Der Heyden F H J, Bonthuis D J, Stein D, Meyer C and Dekker C 2006 Electrokinetic energy conversion efficiency in nanofluidic channels *Nano Lett.* **6** 2232–7
[31] Fu C, Lu X and Yang T 2024 Characterization methods on moisture-enabled power generator: mechanism, parameters and applications *J. Phys. D: Appl. Phys.* **57** 373003
[32] Manjakkal L, Núñez C G, Dang W and Dahiya R 2018 Flexible self-charging supercapacitor based on graphene-Ag-3D graphene foam electrodes *Nano Energy* **51** 604–12
[33] Krishnamoorthy K, Pazhamalai P, Mariappan V K, Nardekar S S, Sahoo S and Kim S J 2020 Probing the energy conversion process in piezoelectric-driven electrochemical self-charging supercapacitor power cell using piezoelectrochemical spectroscopy *Nat. Commun.* **11** 1–11
[34] Kil H J, Kim S R and Park J W 2022 A self-charging supercapacitor for a patch-type glucose sensor *ACS Appl. Mater. Interfaces* **14** 3838–48
[35] Xiong W *et al* 2019 A wearable system based on core–shell structured peptide-Co9S8 supercapacitor and triboelectric nanogenerator *Nano Energy* **66** 104149
[36] Zhou C, Zhang Y, Li Y and Liu J 2013 Construction of high-capacitance 3D CoO@polypyrrole nanowire array electrode for aqueous asymmetric supercapacitor cheng *Nano Lett.* **13** 2078–85
[37] Min S, Zhao C, Zhang Z, Chen G, Qian X and Guo Z 2015 Synthesis of $Ni(OH)_2$/RGO pseudocomposite on nickel foam for supercapacitors with superior performance *J. Mater. Chem.* A **3** 3641–50

[38] Lee S W, Kim J, Chen S, Hammond P T and Shao-Horn Y 2010 Carbon nanotube/manganese oxide ultrathin film electrodes for electrochemical capacitors *ACS Nano* **4** 3889–96

[39] Pang H, Abdalla A M, Sahu R P, Duan Y and Puri I K 2018 Low-temperature synthesis of manganese oxide–carbon nanotube-enhanced microwave-absorbing nanocomposites *J. Mater. Sci.* **53** 16288–302

[40] Qin S, Zhang Q, Yang X, Liu M, Sun Q and Wang Z L 2018 Hybrid Piezo/triboelectric-driven self-charging electrochromic supercapacitor power package *Adv. Energy Mater.* **8** 1–9

[41] Panwar V, Chauhan P S, Kumar S, Tripathi R and Misra A 2023 Gate field induced extraordinary energy storage in MoS_2-graphene-based ultramicro-electrochemical capacitor *ACS Energy Lett.* **8** 1510–9

[42] Poonam , Sharma K, Arora A and Tripathi S K 2019 Review of supercapacitors: materials and devices *J. Energy Storage* **21** 801–25

IOP Publishing

Chapter 9

Applications of supercapacitor

Supercapacitors (SCs) are extensively utilized in diverse applications because of their remarkable power density compared to that of batteries. The limited lifespan of batteries and their toxicity limit their applications in various fields, such as medicine. The lower energy density associated with SCs suggests their use as supplementary energy storage systems for many applications. This chapter highlights a few main applications of SCs.

9.1 Electrical vehicles

Manufacturers of electric vehicles (EVs) employ advanced techniques to enhance both the range and overall performance of their vehicles. The distinct advantages of SCs, including their superior power density and extended cycle life, enable efficient recovery of regenerative energy in EVs. Additionally, SCs provide rapid power delivery with reduced fluctuations in the direct current (DC) bus, alleviating strain on the battery system [1].

9.2 Integrated EV systems with supercapacitors and a battery pack

Figure 9.1 is a schematic representation of the integration of an SC with batteries to power the drivetrain of an EV. The accelerator provides the required torque and speed to the motor. During braking, the motor operates to reverse the power flow to recharge the battery and the SC, as illustrated in [1]. The SC can be connected to the battery pack and the EV's DC link using one of three configurations: passive, semiactive, or active, depending on the power demand and battery type [2]. In the passive configuration, the battery and SC are directly connected to the DC link in a parallel connection, bypassing the use of intermediate DC–DC converters [3]. While this approach is both cost-effective and straightforward, it has certain drawbacks. For instance, the terminal voltage is constrained by the discharge profile of the battery, restricting the effective utilization of the voltage of the SC. Additionally,

doi:10.1088/978-0-7503-5944-3ch9

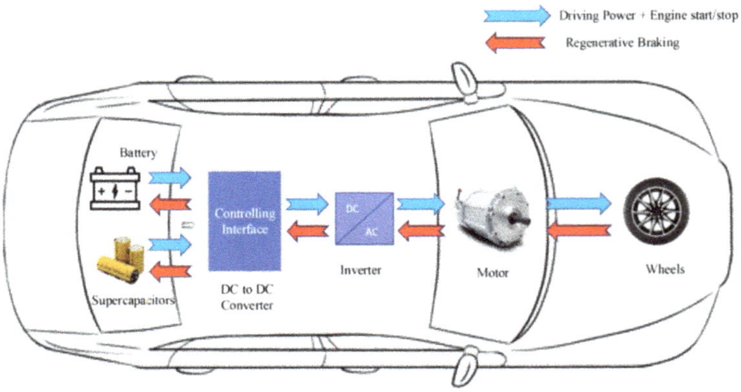

Figure 9.1. Electric vehicle power management. Reproduced from [1]. CC BY 4.0.

once SC discharges, the battery must simultaneously power the load and recharge the SC, which places extra stress on the battery [1, 4]. To address the limitations of passive configurations, the semiactive configuration was introduced. In this setup, the battery is directly linked to the DC bus, while the SC is interfaced with the DC bus via a DC-to-DC converter. The direct connection of the battery allows it to stabilize the DC bus voltage, while the SC is utilized to manage significant current variations [5]. This configuration is often favored due to its cost efficiency and compact design, particularly when a smaller SC pack is employed. However, the direct connection of the battery may result in scenarios where large currents are drawn from it instead of the SC, primarily owing to delays in the converter [6].

The active configuration represents the latest advancement in SC-battery hybrid systems for EVs. In this setup, the SC and the battery are individually connected in parallel to the DC link via separate DC-to-DC converters. This approach enables the SC to operate over a broader voltage range while ensuring a steady current for the battery. A series connection can also be established by adjusting the active configuration [7]. Despite the benefits of this innovative arrangement, certain technical hurdles remain unresolved. For instance, this design requires multiple input parameters and a sophisticated control framework to achieve precise regulation [8]. These setups are advantageous for regenerative braking (RB) and start-stop operations in electric and hybrid vehicles, as they deliver a high burst of power within a short duration.

9.3 Regenerative braking

A common method in contemporary electric and hybrid EVs involves converting kinetic energy into electrical energy during braking by utilizing the motor as a generator, as illustrated in figure 9.1. The reversed power flow is directed toward the SC and the battery. The RB technique enhances the driving range of EVs and the fuel efficiency of hybrid EVs [1]. However, channeling the rapid energy recovery from RB directly into the battery can impose significant stress, leading to accelerated degradation of its long-term performance [9]. To address this, SCs are employed as an effective alternative for capturing regenerative energy, as they are particularly suited for scenarios involving rapid high-power fluctuations. Using the configurations described in the previous section, regenerative energy can be partially or fully diverted into SCs, based on specific requirements. Many leading automobile manufacturers incorporate SCs for energy capture during RB, with examples including the Mazda 6 and 3, Toyota Supra and FCHV, Chevrolet Equinox, and Hyundai Tucson [10]. Notably, a project in Ningbo, China, demonstrated that integrating RB into a trolley bus system equipped with a 1500 V SC pack increased the vehicle's driving range from 4 to 10 km [11].

Numerous studies have explored optimal strategies to maximize the recovery of RB energy without compromising safety systems such as antilock braking systems (ABSs). For instance, one research team proposed an optimization method to determine the braking power needed to maximize energy recovery while maintaining safety. They employed an optimal control framework based on Pontryagin's maximum principle [12]. A method for controlling constant current has been proposed to capture regenerative energy through a passive series configuration, allowing the SCs to exclusively collect the energy while reducing the burden on the battery [9]. A simulation study evaluated the energy recovery efficiency of $LiFePO_4$ batteries compared to SCs [13]. Its findings indicate that $LiFePO_4$ batteries exhibit a recovery efficiency of around 67%, whereas SCs achieve an efficiency of 97% in capturing regenerative energy.

A new method has been introduced for regulating the regenerative power transfer between an SC and a battery [14]. In this approach, the controller directs the current to either the battery or the SC during RB based on predefined current thresholds. Subsequently, current is shared between the battery and the SC within their limits to maintain charge balance. However, a limitation of this method is its reliance on fixed current thresholds, which can vary across systems. Additionally, as batteries age and their current ratings shift, these thresholds require periodic updates.

9.4 Start-stop systems

When restarting EVs or hybrid EVs, a substantial current is typically required for a brief duration. Extracting this current from a conventional battery imposes considerable strain, leading to a reduction in its lifespan. SCs, with their lower equivalent series resistance (ESR) compared to that of batteries, are well suited to deliver this burst of power, facilitating vehicle startup [15]. This approach not only minimizes stress on the battery, prolonging its lifespan, but also ensures quicker and smoother startups [15, 16]. Additionally, SCs are a favorable choice due to their maintenance-free nature and durability, often matching the vehicle's lifespan [16]. Maxwell Technologies, in collaboration with Continental Automotive Systems, has developed a voltage stabilization system powered by Maxwell ultracapacitors, which are widely utilized in automotive applications for engine start-stop functionality. This system is implemented in Cadillac CTS sedans and Cadillac ATS sedans and coupés [16].

The SkelStart Engine Start Module made by Skeleton Technologies is an advanced SC solution designed for trucks, primarily used to optimize the start-stop function. Operating at 12 V with a capacitance of 1280 F, it delivers a high discharge current of 7644 A in just 0.4 s and achieves a peak power output of 62.9 kW. With an operating temperature range from −40 °C to +65 °C, the module is suitable for a wide variety of environmental conditions [17]. Installed between the battery and starter motor of the vehicle, it powers the starter motor, while the battery supports other electrical systems, such as lights and fans. Studies indicate that this module has the potential to reduce fuel consumption by 2%–6% while increasing battery lifespan by 1.5–2 times [17].

9.5 Uninterruptible power supplies

SCs are energy storage devices that provide short-term power backup in uninterruptible power supply systems, offering advantages such as high power density, rapid charging and discharging, long lifespan, and low maintenance [18]. While they cannot replace batteries due to their lower energy density, they are ideal for bridging power gaps during outages or generator startups. Despite higher initial costs, long lifespan of an SC and reliability may offset this, making expert consultation essential for proper integration into UPS systems [19, 20].

9.6 Microgrid applications

Microgrids have recently gained popularity due to their use of distributed power generation. Many of the power sources in microgrids, such as solar and wind, are highly dependent on weather conditions and are considered intermittent. As a result, energy storage is necessary to address fluctuations in power supply. While batteries are commonly employed for this purpose, their performance can deteriorate, and their lifespan can be shortened by rapid charge and discharge cycles [2, 21]. SCs offer an ideal solution to stabilize power fluctuations and reduce strain on batteries. Consequently, hybrid energy storage (HES) systems, which combine batteries and SCs, are often used in microgrids [2, 21, 22].

Researchers are exploring new approaches to integrate SCs into HES systems. The literature outlines three primary topologies for managing HES systems: passive, semiactive, and active topologies [23]. In the passive configuration, the inverter is directly connected to both the batteries and SCs [24]. While this method is cost-effective and straightforward, it has several limitations, including reduced control over power distribution between the battery and SC, potential battery damage due to rapid power fluctuations, and the need for precise voltage matching with the DC bus [25]. To address these limitations, the semiactive topology was developed, offering more flexibility in power flow regulation [26]. In this configuration, the battery is connected directly to the inverter, whereas the SC is integrated through a DC–DC converter. This configuration allows the battery to stabilize the DC link voltage while the SC compensates for sudden variations in power. However, the semiactive topology requires costly, high-current-rated DC-to-DC converters to handle high power demands. The active topology connects both the SC and the battery to an inverter through separate DC-to-DC converters [27]. While it provides superior control over both power distribution and voltage regulation, this approach is the most expensive of the three.

Despite significant research advancements, the practical implementation of SCs in microgrids remains limited, as most HES technologies are still in the research phase. In 2016, a 1.2 MW HES system featuring fast-response ultracapacitors from Maxwell Technologies was implemented by Duke Energy in North Carolina, USA [1]. Two years later, in 2018, the Fraunhofer Institute for Solar Energy Systems launched a renewable-energy-powered HES system on Borkum Island, Germany. This HES microgrid incorporates Maxwell ultracapacitors with capacities of 1 MW and 10 kWh alongside Li-ion batteries rated at 500 kW and 500 kWh [1]. In 2019, a successful pilot project involving a hydrogen-SC-based rural microgrid was carried out in the Malaysian village of Orang Asli, located in the mountains. This microgrid combined 4.7 kW of peak solar power, a 1.1 kW fuel cell, and 10 kWh of SCs [28]. As research continues to advance, the practical applications of SCs in microgrids are anticipated to grow significantly in the upcoming years.

9.7 Portable applications

Camera flashes typically demand a high burst of power, with traditional devices relying on batteries to meet this need. This dependency often results in designs that are bulkier, heavier, and more costly. In contemporary flash technology for cameras, SCs are integrated alongside batteries to enhance longevity, simplify design, and boost both the intensity and stability of the flash. Several commercial products, including the Avaett D60 (without battery), Dual 1080P Dash Cam, and SC WIFI GPS Dash Cam-Q6S4, have already adopted this technology.

A camera flash system has been designed based on a dual-cell SC configuration [1, 29]. The system is triggered by various external commands, including activation of the charge pump circuit to charge the SCs, enabling the light-emitting diodes, selecting the flash torch mode, and limiting the flash current. When the charge pump is initiated, the SCs are charged, and upon activation of the flash drive, the supercapacitors release a large amount of power to the flash. To prevent overvoltage conditions, balancing resistors maintain the voltage across each cell at half of the output voltage. To reduce power losses and minimize self-discharge, some flashlight driver chip manufacturers employ active balancing circuits in place of resistors [29]. The current-switching solid-state circuit provides the proper light output for high-quality photography, while also providing current limiting, overload protection, and stable flash performance. SCs are also utilized in cordless screwdrivers to enhance power output and enable rapid charging. For instance, the 5.4 V Flash Cell cordless screwdriver from Coleman offers fast charging (typically within 90 s) and can drive 22 screws per charge [1]. In contrast, a comparable battery-powered screwdriver requires several hours to recharge and can drive 37 screws on a single charge [1]. SCs are advanced energy storage systems that provide several beneficial characteristics, such as high capacitance, elevated power density, and long cycle lifetimes. These devices are increasingly being utilized across diverse applications, including EVs, microgrids, and portable electronic devices.

9.8 Supercapacitors in telecommunications

In telecommunications, achieving high operational reliability often relies on hot standby systems, such as battery backups, parallel UPS units, and redundant DC–DC converters. However, these standby units consume power even when idle, posing a challenge for power conservation, which is a critical concern [18]. To address this, cold standby solutions such as diesel generators and fuel cells are utilized, although they have longer response times [18, 30]. A viable alternative is the use of SCs, which can quickly activate to handle system interruptions.

9.9 Cold-start behavior of diesel-fueled engines

Starting a diesel engine in extremely cold conditions, such as temperatures as low as −40° C, is particularly challenging [30, 31]. Lead-acid batteries, commonly used for engine cranking, encounter increased internal resistance at low temperatures, which significantly hinders their ability to deliver the high current required [18, 30]. As the battery ages, its internal resistance rises further, diminishing its current discharge capacity even more. Incorporating an SC bank alongside the battery can help provide the additional cranking current needed by the motor.

9.10 Applications of microsupercapacitors

In practical settings, in-plane microsupercapacitors (MSCs) function as versatile energy storage devices, either as independent power sources or in conjunction with energy harvesters to support a variety of Internet of Things (IoT) applications [32]. These devices excel in enhancing functionality and integration for IoT systems [33–35]. Key application areas include energy storage, self-sustaining wireless sensors such as gas and ultraviolet (UV) detectors, stimuli-responsive and photoresponsive systems, and AC line filtering. Moreover, MSCs can be configured in parallel or series arrangements to tailor their capacity, voltage, and current output to meet specific application requirements [32]. A detailed exploration of these applications is provided in the next section.

9.10.1 Energy storage devices

Planar MSCs serve as efficient energy storage devices that can be directly paired with energy-harvesting systems [36]. Triboelectric nanogenerators (TENGs) are capable of capturing ambient mechanical energy by leveraging the combined principles of electrostatic induction and contact electrification. Jiang *et al* demonstrated the integration of TENGs with MSCs, creating a system that efficiently transforms mechanical energy into electrical energy, functioning as a unified harvester and storage module [37]. Electricity can be produced when a TENG device interacts with human skin, subsequently charging the MSCs. Additionally, the silicone rubber-based prosthetic skin, reinforced with solid electrolytes, is biocompatible and suitable for wearable or implantable sensors. Furthermore, Maeng *et al* demonstrated the integration of MSCs within a radio-frequency (RF) energy-harvesting system on a single parylene platform [38]. This system efficiently converts RF input into DC while storing energy in MSCs. This system replicates the operating conditions anticipated for autonomous RF biomedical platforms designed for implantable or wearable applications powered by remote RF energy. Notably, this diverse RF energy-harvesting system can be constructed on a single wafer-level, flexible parylene-based framework. The integrated piezoelectric generator (PEG) energy-harvesting system, coupled with MG-MSCs for line filtering, offers a viable replacement for conventional electrolytic capacitors in AC line filtering [32, 39].

9.10.2 Self-powered wireless sensors

Yun *et al* presented a flexible sensor system built on paper, combining a UV sensor with asymmetric microsupercapacitors (AMSCs) featuring a MnO_2 nanoball as the positive electrode and V_2O_5 as the negative electrode [32, 40]. By employing the liquid metal Galinstan for interconnections, the researchers integrated these asymmetric MSCs onto a waterproof mineral paper, demonstrating remarkable mechanical durability [32].

Cai *et al* introduced a self-powered photodetection system based on ZnO nanoparticles by employing direct laser writing on a polyimide film [41]. This system integrates MSCs as the energy storage component with a commercial solar panel, enabling both the conversion and storage of solar energy. Consequently, the self-powered MSC platform facilitates the monitoring of periodic ultraviolet light responses during illumination cycles [32].

The self-discharge characteristics of MSCs play a crucial role in their suitability for sensor networks [32]. When MSCs are charged, they exist in a high-energy state relative to their discharged condition. This energy disparity creates a thermodynamic impetus, driving the phenomenon of self-discharge. During the self-discharge process, a minor leakage current results in a gradual voltage decrease over time. To quantify this leakage current, a rated DC voltage is applied to the MSCs, and the current necessary to sustain this voltage is measured. Considering this factor, Hota *et al* developed integrated devices that demonstrate a gradual self-discharge rate of 18.75 mV h, showing significant potential for use in self-powered systems [42].

9.10.3 Stimuli-responsive devices

Stimuli-responsive MSCs have recently garnered significant attention due to their potential applications in advanced electronic systems. In a recent study, Liu *et al* developed photoresponsive MSCs on a sapphire substrate, incorporating the photochromic diarylethene derivative DAE (1,2-bis(2,4-dimethyl-5-phenyl-3-thienyl)-3,3,4,4,5,5-hexafluoro-1 cyclopentene) [43]. These photo-switchable molecules undergo efficient interconversion between two stable photoisomers when exposed to UV or white light. Additionally, the reversible control of MSCs is ascribed to the photoisomerization process that alters the charge equilibrium at the DAE/graphene interface [32]. The transition between colorless and colored states can be achieved during the charge–discharge cycle by introducing electrochromic additives into the electrolyte. Zhang *et al* developed flexible MSCs based on exfoliated graphene/V_2O_5, where a PVA/LiCl gel electrolyte with viologen was drop-cast onto the interdigital electrode surface [44]. The charge–discharge cycles (0–1.0 V) reveal *in situ* reversible electrochromic behavior, providing a clear indication of the charge and discharge states of the fabricated MSCs.

The phenomenon was ascribed to the equilibrium reaction between the radical ion of viologen (MV^{+}, purple) and its neutral form (MV^{2+}, colorless). As a result, the developed MSCs demonstrate a voltage-dependent electrochromic response.

9.10.4 Medical diagnostics

The increasing demand for painless diagnostic methods drives the trend toward miniaturizing medical devices. However, the unpredictable pathological properties of many biocompatible materials, combined with the discontinuity of electrode components under complex and dynamic surface strain, often result in uncertain performance and significant safety concerns in the diagnostic process. Due to their compact size, excellent flexibility, and high selectivity of electrode materials, MSCs present a promising option for medical auxiliary energy applications [45]. For instance, Gao *et al* designed an edible MSC (EMSC) based on food ingredients and a template-imprinting technique as the production process [46]. The electrode and electrolyte materials used in this structure were biocompatible, nontoxic, and capable of being absorbed or broken down by the human body. The developed EMSC demonstrated outstanding mechanical stability and flexibility, allowing it to conform to objects of various shapes, including human skin and fruit. Additionally, this microdevice can be wound into the casing of a medical capsule, enabling power supply for capsule endoscopy in a simulated gastric juice environment [45]. The key characteristics of MSCs open up opportunities for real-time *in vivo* detection and other potential biological applications. A key category of wearable electronic devices for personal health is flexible, attachable smart systems for monitoring body fluids. These systems enable real-time health status prediction by analyzing the composition of body fluids. Due to the flexibility and selectivity of the substrate, planar MSCs are well suited to meet this requirement. For instance, Lu *et al* developed a perspiration monitoring system that incorporated a $NiCo_2O_4$-based MSC [47]. The fabricated MSC exhibited an energy density of 0.64 $\mu W\ cm^{-2}$. Additionally, the system demonstrated impressive performance in monitoring, showing high sensitivity to ions and glucose in sweat [45].

This system has the potential to be integrated with wireless sensor technology, enabling precise real-time sweat monitoring and transmitting essential data to mobile devices for health evaluation. Additionally, flexible MSCs offer a solution to the common issue of frequent battery replacements caused by the limited lifespans of microbatteries [45].

Furthermore, MSCs derived from biodegradable materials are harmless to organisms. Overall, a thoughtful application of MSCs is anticipated to accelerate advancements and innovations in the field of medicine [45].

9.10.5 AC line filtering

Aluminum electrolytic capacitors (AECs) are extensively utilized in ac line filtering but typically exhibit a large physical size and low volumetric energy density, making them unsuitable for advancing intelligent systems, particularly in on-chip technologies [48]. Compared to AECs, MSCs are compact, easily integrable, and exhibit high power density, making them ideal for ac line-filtering applications. Yang *et al* developed an MSC using an azulene-bridged coordination polymer framework synthesized through a layer-by-layer approach [49]. The fabricated MSC exhibited a maximum specific capacitance of 34.1 $F\,cm^{-3}$ at a scan rate of 50 $mV\,s^{-1}$ and achieved a phase angle of $-73°$ at 120 Hz, with a resistance–capacitance time constant of approximately 0.83 ms. However, limited rate capability continues to be a significant challenge hindering the use of MSCs in filtering applications. To tackle this challenge, MXene and graphene are considered promising materials due to their abundant ionic pathways and rapid surface redox processes. For instance, Xu *et al* demonstrated a high-frequency MSC utilizing a hybrid electrode structure composed of 2D pseudocapacitive MXene and multi-walled carbon nanotubes (MWCNTs) [50]. The MSC demonstrated superior power density and enhanced frequency response at 120 Hz compared to commercial tantalum capacitors, benefiting from MXene's high capacitance as the active material and MWCNT's efficient ion transport pathways as interlayer support. The MSC demonstrated enhanced functionality and compactness compared to electrolytic capacitors when utilized in both low-pass filter and relaxation oscillator circuits [45]. The system effectively powered a light-emitting diode array using an ac voltage. Similarly, Jiang *et al* fabricated a metal–semiconductor–carbon device employing solution-processable 2D MXene nanosheets and successfully utilized it for ac line filtering [51]. Fine-tuning of the flake size, electrode thickness, and the distance between the electrode fingers allowed the fabricated MSC to attain a volumetric capacitance of 30 $F\,cm^{-3}$ at 120 Hz, coupled with a relaxation time constant of 0.45 ms. This performance surpassed that of electrolytic capacitors ($t_0 = 0.8$ ms), making the device suitable for filtering 120 Hz ripples [45]. In summary, MSCs, with their distinctive advantages, show significant promise as ac line filters for electronic devices and the internet of things, effectively filtering 120 Hz ripples [45].

References

[1] Subasinghage K, Gunawardane K, Padmawansa N, Kularatna N and Moradian M 2022 Modern supercapacitors technologies and their applicability in mature electrical engineering applications *Energies* **15** 7752

[2] Jing W L, Lai C H, Wong W S H and Wong M L D 2016 Cost analysis of battery–supercapacitor hybrid energy storage system for standalone PV systems *IET Conf. Publ.*

[3] Kuperman A, Aharon I, Malki S and Kara A 2013 Design of a semiactive battery–ultracapacitor hybrid energy source *IEEE Trans. Power Electron.* **28** 806–15

[4] Gao L, Dougal R A and Liu S 2005 Power enhancement of an actively controlled battery/ultracapacitor hybrid *IEEE Trans. Power Electron.* **20** 236–43

[5] Girard H L, Wang H, D'Entremont A L and Pilon L 2015 Enhancing Faradaic charge storage contribution in hybrid pseudocapacitors *Electrochim. Acta* **182** 639–51

[6] Lukic S M, Cao J, Bansal R C, Rodriguez F and Emadi A 2008 Energy storage systems for automotive applications *IEEE Trans. Ind. Electron.* **55** 2258–67

[7] Kuperman A and Aharon I 2011 Battery–ultracapacitor hybrids for pulsed current loads: a review *Renew. Sustain. Energy Rev.* **15** 981–92

[8] Tu C H and Emadi A 2012 A novel series-parallel reconfigurable hybrid energy storage system for electrified vehicles *2012 IEEE Transp. Electrif. Conf. Expo, ITEC 2012*

[9] Pan C, Wang J, Dai W, Chen L and Chen L 2020 Constant current control for regenerative braking of passive series hybrid power system *Int. Trans. Electr. Energy Syst.* **30** 1–19

[10] Horn M, MacLeod J, Liu M, Webb J and Motta N 2019 Supercapacitors: a new source of power for electric cars? *Econ. Anal. Policy.* **61** 93–103

[11] Lin J, Niu J, Li H and Atiquzzaman M 2019 A secure and efficient location-based service scheme for smart transportation *Futur. Gener. Comput. Syst.* **92** 694–704

[12] Parvini Y and Vahidi A 2012 Optimal charging of ultracapacitors during regenerative braking *2012 IEEE Int. Electr. Veh. Conf. IEVC 2012*

[13] Perrotta D, Ribeiro B, Rossetti R J F and Afonso J L 2012 On the potential of regenerative braking of electric buses as a function of their itinerary *Procedia—Soc. Behav. Sci.* **54** 1156–67

[14] Jin L Q, Zheng Y, Li J H and Liu Y L 2015 A study of novel regenerative braking system based on supercapacitor for electric vehicle driven by in-wheel motors *Adv. Mech. Eng.* **7** 1–12

[15] Xie D, Liu M, Xu L and Lu W 2021 Bi-level programming approach for coordinated configuration of distributed generations and automation devices in distribution networks *Int. J. Electr. Power Energy Syst.* **133** 107210

[16] Taranovich S 2015 https://www.edn.com/general-motors-chooses-ultracapacitors-for-start-stop/

[17] Skeleton Technologies 2025 https://www.skeletontech.com/skelstart-ultracapacitor-engine-start-module

[18] Gautham Prasad G, Shetty N, Thakur S, Rakshitha and Bommegowda K B 2019 Supercapacitor technology and its applications: a review *IOP Conf. Ser.: Mater. Sci. Eng.* **561** 012105

[19] Wayu M 2021 Manganese oxide carbon-based nanocomposite in energy storage applications *Solids* **2** 232–48

[20] Lahyani A, Venet P, Guermazi A and Troudi A 2013 Battery/supercapacitors combination in uninterruptible power supply (UPS) *IEEE Trans. Power Electron.* **28** 1509–22

[21] Şahin M E and Blaabjerg F 2020 A hybrid PV-battery/supercapacitor system and a basic active power control proposal in MATLAB/simulink *Electronics* **9** 129

[22] Glavin M E, Chan P K W, Armstrong S and Hurley W G 2008 A stand-alone photovoltaic supercapacitor battery hybrid energy storage system *2008 13th Int. Power Electron. Motion Control Conf. EPE-PEMC 2008* 1688–95

[23] Babu T S, Vasudevan K R, Ramachandaramurthy V K, Sani S B, Chemud S and Lajim R M 2020 A comprehensive review of hybrid energy storage systems: converter topologies, control strategies and future prospects *IEEE Access* **8** 148702–21

[24] Zheng J P, Jow T R and Ding M S 2001 Hybrid power sources for pulsed current applications *IEEE Trans. Aerosp. Electron. Syst.* **37** 288–92

[25] Ma T, Yang H and Lu L 2015 Development of hybrid battery–supercapacitor energy storage for remote area renewable energy systems *Appl. Energy* **153** 56–62

[26] Al Takrouri M, Ayob S M, Idris N R N, Aziz M J A, Ayop R, Ghith E, Tlija M, Majeed A H and Arfeen Z A 2024 A comparison study of hybrid energy storage system topologies for electric vehicles *IEEE Access* **274** 400–11

[27] Cohen I J, Wetz D A, Heinzel J M and Dong Q 2015 Design and characterization of an actively controlled hybrid energy storage module for high-rate directed energy applications *IEEE Trans. Plasma Sci.* **43** 1427–33

[28] Enapter S.r.l. 2019 https://www.enapter.com/application/hydrogen-supercapacitor-rural-microgrid/#28958

[29] Texas Instruments 2013 https://www.mouser.com/ds/2/405/snvs569b-261655.pdf?srsltid=AfmBOoquWH1Ync9XTVpFY94ByGhvZJUvsDPlEdb8fuFjK2mstaVD1ell

[30] Shah V, Joshi J A, Maheshwari R and Roy R 2008 Review of ultracapacitor technology and its applications *15th Natl. Power Syst. Conf. (NPSC) (IIT Bombay)* pp 142–7 http://ee.iitb.ac.in/~npsc2008/NPSC_CD/Data/Poster/p254.pdf

[31] Pay S and Baghzouz Y 2003 Effectiveness of battery–supercapacitor combination in electric vehicles *2003 IEEE Bol. PowerTech—Conf. Proc. 3* pp 728–33

[32] Zhang J, Zhang G, Zhou T and Sun S 2020 Recent developments of planar micro-supercapacitors : fabrication, properties, and applications**30** 1910000

[33] Pu X, Hu W and Wang Z L 2018 Toward wearable self-charging power systems: the integration of energy-harvesting and storage devices *Small.* **14** 1–19

[34] Luo J, Fan F R, Jiang T, Wang Z, Tang W, Zhang C, Liu M, Cao G and Wang Z L 2015 Integration of micro-supercapacitors with triboelectric nanogenerators for a flexible self-charging power unit *Nano Res.* **8** 3934–43

[35] Kurra N, Jiang Q, Syed A, Xia C and Alshareef H N 2016 Micro-pseudocapacitors with electroactive polymer electrodes: toward AC-line filtering applications *ACS Appl. Mater. Interfaces* **8** 12748–55

[36] Qiu M, Sun P, Cui G, Tong Y and Mai W 2019 A flexible microsupercapacitor with integral photocatalytic fuel cell for self-charging *ACS Nano* **13** 8246–55

[37] Jiang Q, Wu C, Wang Z, Wang A C, He J H, Wang Z L and Alshareef H N 2018 MXene electrochemical microsupercapacitor integrated with triboelectric nanogenerator as a wearable self-charging power unit *Nano Energy* **45** 266–72

[38] Maeng J, Meng C and Irazoqui P P 2015 Wafer-scale integrated micro-supercapacitors on an ultrathin and highly flexible biomedical platform *Biomed. Microdevices* **17** 7

[39] Ye J *et al* 2018 Direct laser writing of graphene made from chemical vapor deposition for flexible, integratable micro-supercapacitors with ultrahigh power output *Adv. Mater.* **30** 1–8

[40] Yun J, Lim Y, Lee H, Lee G, Park H, Hong S Y, Jin S W, Lee Y H, Lee S S and Ha J S 2017 A patterned Ggraphene/ZnO UV sensor driven by integrated asymmetric micro-supercapacitors on a liquid metal patterned foldable paper *Adv. Funct. Mater.* **27** 1–13

[41] Cai J, Lv C and Watanabe A 2016 Laser direct writing of high-performance flexible all-solid-state carbon micro-supercapacitors for an on-chip self-powered photodetection system *Nano Energy.* **30** 790–800

[42] Hota M K, Jiang Q, Wang Z, Wang Z L, Salama K N and Alshareef H N 2019 Integration of electrochemical microsupercapacitors with thin film electronics for on-chip energy storage *Adv. Mater.* **31** 1–9

[43] Liu Z, Wang H I, Narita A, Chen Q, Mics Z, Turchinovich D, Kläui M, Bonn M and Müllen K 2017 Photoswitchable micro-supercapacitor based on a diarylethene-graphene composite film *J. Am. Chem. Soc.* **139** 9443–6
[44] Zhang P, Zhu F, Wang F, Wang J, Dong R, Zhuang X, Schmidt O G and Feng X 2017 Stimulus-responsive micro-supercapacitors with ultrahigh energy density and reversible electrochromic window *Adv. Mater.* **29** 1
[45] Bu F, Zhou W, Xu Y, Du Y, Guan C and Huang W 2020 Recent developments of advanced micro-supercapacitors: design, fabrication and applications *Npj Flex. Electron.* **4** 31
[46] Gao C, Bai C, Gao J, Xiao Y, Han Y, Shaista A, Zhao Y and Qu L 2020 A directly swallowable and ingestible micro-supercapacitor *J. Mater. Chem.* A **8** 4055–61
[47] Lu Y, Jiang K, Chen D and Shen G 2019 Wearable sweat monitoring system with integrated micro-supercapacitors *Nano Energy.* **58** 624–32
[48] Zhao D *et al* 2019 Charge transfer salt and graphene heterostructure-based micro-supercapacitors with alternating current line-filtering performance *Small* **15** 1–9
[49] Yang C *et al* 2017 Coordination polymer framework based on-chip micro-supercapacitors with AC line-filtering performance *Angew. Chemie—Int. Ed.* **56** 3920–4
[50] Xu S, Liu W, Hu B and Wang X 2019 Circuit-integratable high-frequency micro supercapacitors with filter/oscillator demonstrations *Nano Energy.* **58** 803–10
[51] Jiang Q, Kurra N, Maleski K, Lei Y, Liang H, Zhang Y, Gogotsi Y and Alshareef H N 2019 On-chip MXene microsupercapacitors for AC-line filtering applications *Adv. Energy Mater.* **9** 1–9

www.ingramcontent.com/pod-product-compliance
Ingram Content Group UK Ltd.
Pitfield, Milton Keynes, MK11 3LW, UK
UKHW051851100926
8556IPUK00005B/567

9 780750 359429